PLATO'S CAVE
Revised Edition
Television and its Discontents

CRITICAL BODIES
Joseph Pilotta, series editor

Plato's Cave
  Revised Edition
  Television and its Discontents
    John O'Neill

Knowledge and the Production of Non-knowledge
  An Exploration of Alien Mythology in Post-War America
    Mark Featherstone

forthcoming

The Body in Human Inquiry: Interdisciplinary Explorations of Embodiment
    Vicente Berdayes, Luigi Esposito and John M. Murphy

Visceral Manifestation and the East Asian Communicative Body
    Jay Goulding

The New Age Ethic and the Spirit of Postmodernity
    Carmen Kuhling

Hispanic Tele-visions in the United States: Television Discourse and the
    Cultural Construction of Ethnicity
      Elizabeth Lozano

Abjection and its Correction in Ethnic Studies
    Jill McCaughin

The Seductive Aesthetics of Post-Colonialism
    Rekha Menon

Body Works: Essays on Modernity and Morality
    John O'Neill

Why I Want My MTV: Music Video and Aesthetic Communication
    Kevin Williams

PLATO'S CAVE
Revised Edition
Television and its Discontents

John O'Neill

York University
Toronto, Canada

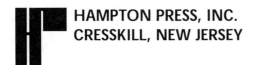

HAMPTON PRESS, INC.
CRESSKILL, NEW JERSEY

Printed in the United States of America

Library of Congress Cataloging-in-Publication Data

O'Neil, John, 1933-
   Plato's cave : television and its discontents / John O'Neill.--Rev. ed.
   p. cm. -- (The Hampton Press communication series. Critical
   bodies)
   Includes bibliographical references and index.
   ISBN 1-57273-389-6 -- ISBN 1-57273-390-X
   1. Mass media. 2. Television. I. Title. II. Series.

P90.O49 2002
302.23'45--dc21

                                                         2001051994

Hampton Press, Inc.
23 Broadway
Cresskill, NJ 07626

CONTENTS

PREFACE: CAPITAL TV

The time of thought is always out of joint. Yet like Hamlet we must fit thought to the action. When all the action appears to be on TV, the time of analysis goes dead. Or so it would seem, unless we realize that like the rest of our fast culture TV is the prime producer of dead time. Thus TV is never as dead as when it attempts to report live events to which it fixes like a vampire in the hope of animating itself. Having no mind and no speech, TV cannot think the events it claims to animate for us. This is because TV has no more time to think than it has to call itself television. How did time get so scarce? It is the nature of *capital time* to speed up our lives so that there is no time-out from the business of business—unless we are sold our relaxation, holidays, and inner calm. We cannot buy-out from capital time because it has already absorbed the aboriginal, the aristocratic, the nostalgic, the revolted and the outrageous in postmortem celebrations of capital culture. Because global capitalism is the absent body of capital authority, TV must fill the political scene/seen with stories of bad government, terrorism, food mutations, health warnings and hurricane advisories that usurp the provision of civic order.

 TV advertises itself. We watch it as voyeurs hoping that our own lives might appear on the screen, whether as a newsworthy victim, an occasional hero or a survivor of someone else's misfortune. Whatever the case, the speechlessness of "being-there" confirms the scene as an occasion for *capital sensation*. In exchange for absorbing citizen response, TV promises to deliver a comprehensive report on the causes of its shock events in the name of the people's "right to know." Thus *capital TV* supplies us with our thoughts and feelings, with our families, friends, communities and world. It does so repeatedly. This is not because of the seriousness of its mission but because TV is always returning from its "commercial breaks" that are the cross we all bear for capital TV.

Why would anyone try to understand TV? Doesn't that mean mis-understanding it—by treating TV as a book, by looking behind the screen rather than at the screen? Isn't the whole exercise of cultural criticism itself a dead enterprise, hopelessly tied to scholarly time-out? The very practice of TV analysis clashes with the *dating complex* in capital time. Thus *Plato's Cave* runs the risk of being already out of date because its references to political and cultural events, especially film and TV series, cannot be up to the minute (it takes time to produce a book). Yet TV's instant history cannot yield historical perspective which we achieve only through framing events as events in relatively stable structures, systems or institutions. Thus what *Plato's Cave* argues is that the detail of any TV show may vary endlessly (and must to fulfill the commercial imperative of continuous entertainment) yet never alter the generic nature of a detective story, a news report or a sci-ence-fiction adventure. In every case, the *god-term* of the social system (law, science, capital, life, sexuality) will be pitted against its *devil-term* (crime, disease, poverty, death, desirelessness) in a narrative that resolves all misadventures in favor of the social order and individual confirmation. In this sense, there is only ever one show on TV because its function is to mir-ror capitalism as a singing, talking, eating, drinking, loving, passionate body like our own. *Each new season merely repeats the (un)natural cycle of TV.*

Of course, TV cannot escape the cultural contradictions of capital-ism. Thus *Ally McBeal* must have Ally go home lonely every night exhausted by trying to combine professional independence with the heart's dependence. Ally can no more "have it all" than any other American who struggles to be successful, wealthy, beautiful and fortunate without recognizing that failure, poverty, ugliness and misfortune are the other side of the capital coin. Ally and her colleagues "know" this but they are left to recognize it only in scenes of infantile rage or in confessional practices that need the toilet as a setting as much as their legal practice needs the courtroom. Such shows (and everyone has a favorite) serve to have us believe that the law, medicine, and the mili-tary are compatible with romance, sexuality and sentimentalism. By contrast the evil in our worst enemies lies in their unemotional and unswerving pro-fessionalism which marks them as aliens rather than Americans (witness any cyberfiction).

Is there any possibility of breaking up capital time rather than suc-
cumb to it in schizophrenic gestures of speeding-up desire, in well-organized
capital concerts of rock, rap and rave? I think that capital time is challenged
only by *messianic time* in which we (re)turn to our senses. As I see it, the
time of the sixties was a brilliant collective epiphany in which we turned
against the capital culture of dead organizational time in business, church
and war. This turn failed because its sexuality and narcosis could not sustain
the body's revolt against death. It was a brief revolution whose resource lay
in the life of the mind that loves its body and never abandons their union.
This is the love story we must preserve into the next millennium, and already
there are signs that we are again ready to protect the global order fever of
faster wealth and ever deeper poverty and crisis.

In *Plato's Cave* I have adopted a double strategy of framing television as both a prop and an implant. We owe this discovery to Marshall
McLuhan's *Understanding Media: The Extensions of Man* (1965). He first
saw television as a body whose potential for global community he espoused
like Catholicism. I have tried to develop McLuhan's vision with more attention to political economy, body politics, and biotechnology. I exploit the
conflict between the capital bodies of TV and the communicative body in
my earlier works, *Five Bodies* (1985), *The Communicative Body* (1989),
Bodyworks (2002) and *Incorporating Cultural Theory* (2001). I will push
forward the work in *Plato's Cave*. Meantime, dear reader, you have work to
do not to be overwhelmed by the rhetoric of novelty, topicality and instanta-
neity which fixes you before television in order to put your mind and vision
out of play. Do try to hang on to the rule that there is only ever one thing on
TV—namely, TV itself. If you ever understand one show you are free to
take or leave the rest. That would give you a genuine channel change. Life
is short, but bad art makes it even shorter.

ABOUT THE AUTHOR

John O'Neill is Distinguished Research Professor of Sociology at York University, Toronto, and an External Associate of the Centre for Comparative Literature at the University of Toronto. He is also a Fellow of the Royal Society of Canada. He is the author of, among other works: *Essaying Montaigne: A Study of the Renaissance Institution of Writing and Reading* (2001); *Five Bodies: The Human Shape of Modern Society* (1985); *The Communicative Body: Studies in Communicative Philosophy, Politics, and Sociology* (1989); *The Missing Child in Liberal Theory* (1996); and the *Poverty of Postmodernism* (1995). He is presently at work on *The Domestic Economy of the Soul*, a study of Freud's principal five case histories.

PART I

POLITICAL ECONOMY AND THE MEDIATION OF DESIRE

Cave Culture and Tele-vision: Opening and Closing the American Mind

CAVE CULTURE AND THE CLOSURE OF THE AMERICAN ACADEMY

I begin with a passage from—or towards—*Plato's Republic*, a book whose terrifying logic and beauty I have loved since my undergraduate days, at the London School of Economics and Political Science, when I read as hard as I could breathe. The *Republic* is beautiful and logical because it is the construction of an orderly universe—a cosmos reflected in all its parts, in the polity as in the soul. It is terrifying because it is severe with those who do not understand the cosmos and whose ignorance necessarily confines them to its lower levels, to a world of shadows and seductive ignorance—like the movies, or television or newspapers from which I was also trying to learn something about such matters. At that time, I could hardly imagine that my reading of the *Republic* would one day return me to the cave culture to attempt some charitable defense of the necessary connection between the cave and the academy.

Consider the original state of things—a condition in which most of us are generally to be found:

Next, said I, here is a parable to illustrate the degrees in which our nature may be enlightened or unenlightened. Imagine the condition of men living in a sort of cavernous chamber underground, with an entrance open to the light and a long passage all down the cave. Here they have been from childhood, chained by the leg and also by the neck, so that they cannot move and can see only what is in front of them, because the chains will not let them turn their heads. At some distance higher up is the light of a fire burning behind them; and between the prisoners and the fire is a track with a parapet built along it, like the screen at a puppet-show, which hides the performers while they show their puppets over the top.

I see, he said.

Now behind this parapet imagine persons carrying along various artificial objects, including figures of men and animals in wood or stone or other materials, which project above the parapet. Naturally, some of these persons will be talking, others silent.

It is a strange picture, he said, and a strange sort of prisoners.

Like ourselves, I replied: for in the first place prisoners so confined would have seen nothing of themselves or of one another, except the shadows thrown by the fire-light on the wall of the Cave facing them, would they?

Not if all their lives they had been prevented from moving their heads.

And they would have seen as little of the objects carried past.

Of course.

Now, if they could talk to one another, would they not suppose that their words referred only to those passing shadows which they saw?

Necessarily.

And suppose their prison had an echo from the wall facing them? When one of the people crossing behind them spoke, they could only suppose that the sound came from the shadow passing before their eyes.

No doubt.

In every way, then, such prisoners would recognize as reality nothing but the shadows of those artificial objects.

Inevitably.[1]

In this passage we have a remarkable allegory of the age of television whose extreme tendencies are identified with postmodern culture and either celebrated or reviled by self-appointed (anointed?) critics in academia. I am one of these and what I have been thinking about these developments is offered to you in *Plato's Cave*, a title I have chosen in order to try to preserve the tension between critical thought and the body politic. I am therefore concerned at the very outset to set forth my differences of opinion on the issues of "higher education" raised so polemically in Allan Bloom's

The Closing of The American Mind,[2] since I think Bloom's Platonism represents one of the most pernicious attacks upon university and mass culture
(and their unavoidable relationship) in recent times.[3] What divides us is
Bloom's pessimism about the possibility of enlightenment.[4] He believes that
the darkness of the cave cannot be dispelled because most men and women
crave the very darkness of their souls. No light can be brought into the cave;
only a few may escape into the outside light of the philosopher's sun.
Modern enlightenment thinkers misunderstand the allegory of the cave, or
the problem of the relation of knowledge to civil society, because they
aspire to the removal of the asymmetry between the wise and the unwise.

As Bloom would have it, the countermyth of the Enlightenment is
the educability of the unwise and the reduction of all knowledge and art to
serve the demigods of mass society. Since Bloom's argument is itself so
painfully clouded in its ignorance of the tradition of critical Marxist and
humanist thought to which it traces the contemporary troubles of the university—its decomposition—all one can do is to ask the reader to see how this
tradition is employed in the present volume on behalf of a universal knowl-
edge-community whose very idea Bloom rejects out of hand. But it is especially important that the reader see that, despite the critical intent of *Plato's
Cave*, it never entirely abandons the necessary follies of mass culture.
Bloom's excoriation of the youth culture entirely misses its target while
shamelessly exonerating the corporate and military institutions to whom he
makes silent appeal on behalf of his lost academy. It is this corporate world
that has turned all students into part-time students on behalf of full-time
consumerism— rendering the corporate image youthful just as it now feminizes and colors its image in a further intensification of consumerism.
Bloom fails to analyze the imagery of mindless youth and boundless energy
that is generated by media capitalism in order to close the American mind to
any understanding of corporate operations. The most significant step here is
Hollywood's remake of the Vietnam war to prove that America did win a
war that the students had lost for America, even if the platoon did get a little
high and deaf on its music and spent most of its time in a homoerotic man-
hunt.

Nothing enrages Bloom like the wrong music. If anything condemns student cave culture, it is its light and sound system which has
flipped Bloom's Platonic switch over to the driving rock and strobe lights
nurturing infantile sexuality and the politics of fascism. Such students are
the perfect bait for "The German connection"—Bloom's woeful story of the
influence of contemporary European thought upon the university he struggles to keep under the restraint of his Platonism. Whereas Bloom considers
that current European thought—of which he is not very knowledgeable—
has turned Plato's cave into a discotheque, I believe that Bloom's insistence

upon "ethnic" goodness reveals nothing more than a Jewish youth's lament that the Wasps have abandoned the Greeks to let in the Blacks—and the rest of us. "Bloomsday" quite exceeds Bloom because his politics were so ignorant of the political economy of exclusion in America. One can hardly believe that he has forgotten that passing through the university was once as hard for him as passing through the eye of heaven's needle is said to be for the idle rich.

What is extraordinarily uncharitable in Bloom's attack upon cave culture is his reluctance to see that, for all its faults, the modern university has broadened the needle-eye of culture. Admittedly, to achieve this we all pay the price of mass education. But with some patience we may achieve a society where fewer are chosen and many more called to learn. In North America, students come from every land on earth. It is no longer easy to homogenize them with a patriarchal, Judaeo Greek culture whose authority is given with its disdain of the cave culture of the masses. This does not mean that teachers are drowning along with their students in a morass of cultural indifference that can only be stemmed by mounting courses anchored in the Great Books. Of course, one can understand Bloom's romance of books—we all began at the breast of the book. But Bloom wants to keep us there—unweaned because at bottom Bloom wants to possess the student soul forever. This is because he wants to beguile them with the music of Strauss—Bloom's own teacher from whom he imbibed the higher secrets of Plato's eternal truth. What goes unnoticed in this vein of Platonism, with its own seductive song of a higher truth that outwears any contingent opinion, is that this truth is reproduced through the caste machine of the *Republic*—which is a long shot from the American republic which the Straussians and Co. seek to return to the right ways of political education.[5] The *Republic* is a machine for freezing history. As such, it cannot possibly offer a model either for our society or for the university, neither of which can indulge the privilege of nostalgia underwritten by a constitutional hermeneutics to which only the Straussians have the key.

When the winds of history make Americans unsure of themselves it is much more flattering to be told by Bloom rather than Batman that the Founding Fathers had grasped in a flash the essential charter that would steer them safely—provided they keep their eye on the rear view mirror of the Constitution. Americans love the game of wandering from the scriptures only to return to them with thousandfold enthusiasm. The Straussians are more flattering than other political evangelists even if a little less all-American than the apple pie upon which the United States is also founded, as I show in the following chapter. It is extraordinary that Bloom's *Closing of the American Mind* should have become an American best-seller. Did mass culture go on holiday or did the American elite decide to come out of the closet? Bloom's message is anti-intellectual and owes its rewards to that very effect. Of course, Bloom's attempt to close the public mind is wrapped up in academic prose which gives his countercultural project a veneer of objective knowledge. The urgency of its spiritual diagnosis invites Americans to declare another war against themselves. But this time it is a war of the elders against the young, of professors against students, of books against music, of reason against passion. Above all, it is a war of the canon according to Bloom against the revolution of a hundred flowers—a war conducted from the ivory tower on the edge of the Chicago ghetto overlooked by Bloom. *The Closing of the American Mind* was a doomsday book ignorant of the wilderness into which it called because it lacked any comprehension of the sensory culture that is marketed by the corporate institutions to whose seriousness it appealed. Bloom not only neglected Socrates' practice of philosophizing in the marketplace but also failed to ask whatever happened to the market—let alone philosophy, to separate them as much as they are nowadays. Bloom missed his target, attacking youth when he should have analyzed the political economy that generates the docile citizenry required by the corporate republic of America, as I show in Chapter Two. Moreover, he was peculiarly deaf to the student protests of the 1960s which, so far from expressing the moral debilitation of cultural relativism, were inspired by its principle of respect for the integrity of other cultures and the insanity of trying to burn them down in a no-name war.

MEDIA AIDS: SOCIAL AMNESIA AND CARNAL IGNORANCE

The following essays take on the issues for which the imagery of Plato's cave represents a challenge to contemporary self-understanding. For we cannot escape into a higher metaphysics or into an elite culture that will exclude more than half of humanity in a self-inflicted wound of pride and privilege. As a social scientist, one must from time to time remind oneself of the central issues that require one's adopted perspective upon human conduct. This exercise is worthless if it forces out of our attention events, values, and beliefs that in everyday life concern us deeply and upon which we expect there to be some constructive scientific analysis. When this happens, common sense knowledge, having been abandoned by science, is driven to rely upon those very forces in the media arts which serve to deepen our ignorance of the world. Thus, by keeping us up to date with its events, the media leave us mired in pseudofactualism and pop-perspectivism which deepen our social amnesia. None of us escapes this assault upon the mind and body. Yet, it is extraordinarily difficult to discipline ourselves in the proper exercise of that analytic and aesthetic thinking which needs to be fed by an imaginative grasp of the ordinary events that determine our lives. Unfortunately, social scientists easily retreat into the reproduction of knowledge whose own organizational requirements bind the profession to the dominant institutions that shape the forces of criticism and docility in their society. Our awareness of this dilemma makes us all post-Enlightenment thinkers. That is, we know now that while knowledge and communication are the mainstay of power, the social accumulation of knowledge is no longer dependent upon a romantic individual drive for freedom. In other words, large institutions also pursue knowledge, monopolize its communication, and employ it in self-serving ways. Power and communication are knowledge, and as such they may work to reduce the romance of knowledge as emancipatory power. Since modern societies are largely knowledge societies, the problem of the communication and analysis of alterative frameworks of the information that creates "publics," "consumers," "clients," and "voters" becomes vital to our civic lives and to the viability of our concept of democracy.

In the following chapters, I employ analytic and descriptive concepts of desire and the body, at one level, and of ideology and of the corporate and state agenda, at another level. I show how these levels interact through what I call the *specular functions of the media*. I include in my concept of the media the social sciences as multipliers in our pursuit of welfare, health, and therapy, as well as of our pursuit of consumerism and defamilism. In modern society we have no knowledge of ourselves that can be extricated from our knowledge of the limits of knowledge and of our own history. This is not just a matter of intellectual modesty. For we are not modest in our claims upon knowledge, nor in the exercise of power over ourselves. How is that? Well, we are not directly opposed to ourselves. Nor do we willingly impose ignorance upon ourselves. It is "our" institutions—our family, our corporation, class, racism, sexism, scientism, and sensationalism that mediate the differences we suffer between expectation and reality, between unaccountable authority and responsible freedom. To make such claims is not to argue for the politicization of knowledge. It is not to reduce science to ideology, or art to propaganda. It is rather to make it clear that *modern politics are both scientized and mediated* and that this is the basic mechanism of our political system inasmuch as it functions largely through *an ideology of specularized self-conscription* rather than through the resort to blatant force and propaganda—even though this very difference is thereby spectacularly minimized. To think, in such a context of knowledge, means that we must have a firm grasp on the central analytic issues in the social sciences at the same time that we try to cultivate an imaginative empiricism in order to spark the connection between cultural theory and political practice. What weakens the link between thought and practice is not the inherent weakness of either element but rather their ideological separation. That is to say, it is "our" political institutions that scramble everyday experience and undermine its moral cogency in the same way that they expropriate the moral relevance of our sciences. It is this polarization that renders reasonable thought pejoratively "academic," despite the fact that the academy easily sells itself to the purposes of the corporate and state agenda of relevance and expediency.

Many readers will dismiss the phenomena I am concerned with as regressive or else purely expressive behavior. In general, they regard the political process as immune to cultural analysis. Any appeal to the poetry of life as the basis of a new politics arouses the cynicism if not the despair of political observers. That is because the reality which dominates our political life is the product of corporate and state practices governed by a conception of rationality that represses any relation to life and nature other than the exercise of domination. This reality is manufactured through the subrationality of science and the institutions, language, and images that foster the assumption that the good life is only achievable through, or else the same thing as, an expansionist practicality.

In order to grasp the everyday sense of these connections, we need to consider technology as revelation, as an instrument of appearances and the construction of truth in the assertions of cathedrals, skyscrapers, noise, and theater. It is then possible to see that the meaning of our everyday lives in the urban industrial world is a construction of what it is we are trying to say, how we are trying to live among telephones, computers, films, television, automobiles, airplanes, refrigerators, and contraceptives.

To envision these things, however, is to rely upon the conversation of machines that reveal the "mechanical" as the forgetfulness of the presence and absence of man in the technological world. Such is the achievement of Charlie Chaplin's films, the theater of Brecht, Marx's *Capital* and the "much-music" of our times. The machine-paced world is revealed through the films of Charlie Chaplin, whose shuffle reverses the scenarios of assembly and falling grace, turning a pratfall into a political act. In this way *capitalism as the life-form of things* is dramatized and the absence of man implicit in Marx's analytic of *Capital* is brought to expression and reversal through art. For this reason the body politic is dependent upon art for revealing evil as the construct of a rule (a machine) for the accumulation of knowledge or of power and the forgetfulness of alternate social orders. We are in search, then, of an *aesthetics of technology* to serve as the basis of the modern connection between communication and politics and the dissolution of the mind-body dualism which determines the orders of reality and illusion in our political life. Our task is to recover cultural imagination from forgetfulness of the "mechanical," which has always served as the metaphor for the complex and remote organization of the social and political institutions that arose with the power of machines.

One of the consequences of the ideological weakening of the bond between analytic theory and common-sense knowledge and values is the prevalence of *social amnesia.* By this I mean that, despite the proliferation of considerable institutions of public education and of the commercial media, we know less and less about "our" social institutions and acquire an even shorter historical perspective upon them. "Our" presence to journalized (day by day) experience removes us all the more surely from its intelligent evaluation. On the individual level, the counterpart of this collective amnesia is *carnal ignorance*, that is, the process whereby our bodies and our families know less and less in favor of the system of consumerized and medicalized advice. In response to the double pressures of social amnesia and carnal ignorance, I have organized the chapters in Part Two to connect with an analysis of the state therapeutic complex that is now designed to recode the life-world and the body politic so as to reduce the political realm to the administration of a docile citizenry, entertained most evenings by the vicarious suffering of "inner city" and "foreign" struggles and now by "twenty-four hour" news coverage! If the pacification of our citizenry were ever fully achieved, we should have lost all need for the sociological tradition that I treat as a mode of historical knowledge whose civic value gives it a claim upon our humanity. To the extent that the body politic can be reduced to the organizational practices of consumerism and medicalism, we will have suppressed our political history—and thereby the future for which it necessarily calls.

The vocabularies of amnesia and ignorance are deeply implanted in the contemporary self-interpretation of our mediated and specularized society. For this very reason, the thought of the classical theorists I have employed in the analysis of contemporary institutions becomes ever more "difficult." All one can do about this is not to increase the *pedagogical nihilism* that rules in our textbooks and journalism by claiming to "simplify" and to "clarify" issues that are as resistant to the mass educated ignorance of our teachers as they are to that of our students. What more one can do is to show how the classical tradition of social and political thought offers a necessary intelligence that is responsive to the institutional setting in which we ought to live with some regard for one another, and for the future inheritability of our institutions. So far as possible, these chapters avoid any dogmatic posture. Where, for example, they treat history and theory, or psychoanalysis and semiotics seriously, they do so for the sake of the analytic understanding they yield. Similarly, where these essays reveal their inspiration in the political and cultural events of the 1960s, they do so in order to reprieve a history whose mediatized suppression renders even more effective the administrative politics of our day. Thus the following chapters are resolutely tied to our everyday contexts of interrogation and practice.

I should also say that these chapters start from a well-grounded sociological fact that class inequality is an identifying feature of industrial societies. All kinds of activities, objects, relations, discourses, and signs are brought to bear on the stratifying processes that reproduce the society and political economy of inequality. *Liberal state capitalism operates not simply as a goods-machine but as an ideology-machine.* That is to say, it reproduces its own order with least disruption, resistance, and demoralization as long as it is able to mobilize acceptance of the distance between the individual's objective class opportunities and his or her subjective aspirations to unequally available opportunities. This means that the ideological-machine must also provide for the foreclosure of a political questioning of the allocative system of goods and values which might envision wholesale redistributive justice rather than marginal transfers of income through taxation, health, education, and welfare. In this system political parties produce redistributive discourses that vary around a central *foreclosure* of the question of radical social justice. This is a large part of the phenomenon of *social amnesia* which, as I see it, functions to reproduce the ideological stability of liberal capitalism and its current postmodern celebration. Thus the political movements of the 1960s remain important for their contribution to the arousal of memory in the body politic. The latter is not simply an exercise in social regression. Rather, it is the lifelong and lifegiving refusal of injustice matched by the determination to work for a measure of happiness. I have tried, therefore, to distill from the political history of the 1960s a model of the *body politic* in which the insistence upon justice in the family, in the workplace, and the market, as well as a reach for personal happiness, can be framed as necessary articulations of our political discourse. I am not romanticizing the 1960s, by any means. But I am opposed to the current political amnesia with respect to this most recent phase of our political history— already ancient history in our universities and totally lost in the mass media except as it lives by recycling the past out of mind.

The media are part of the current social amnesia that characterizes a society in which the political question is foreclosed. They play an essential role in the production of what I call *carnal ignorance* inasmuch as they translate our political ignorance into our desire for opinion and sensation as our rightful demand upon every other institution from the family, the university, and the market to the health and happiness industries of the therapeutic society. Of course, such negative concepts as social amnesia and carnal ignorance are counterintuitive markers for a society that considers itself to stand on a pinnacle of scientific and carnal knowledge. There is a risk that such negative terminology will be accused of cynicism and elitism. But the latter are only the marks of a higher-order academic conformism. None of this can be avoided. What matters is whether one can still voice any hope for that vision of a *charismatic society* in which our differences might be exalted in our service of one another's intelligence, sensibility, and love. Here the darkest shadow is not our opposing political ideologies, nor the weakness of our knowledge. The dark shadow, as I argue in my concluding chapter, is our nuclearism in which our will to abide our differences is erased once and for all. Here, surely, the exterminist discourses that have colonized our intergalactic imagination, while habituating our children to the deformed creatures of stellar war, must represent the zero-point in our political culture.

Bloom to the contrary, it is not to be regretted that our colleagues in "North America" seek to understand their own theoretical tradition in terms of its Continental appropriation. Of course, there is a risk that this will render them doubly dependent upon the deconstructionist and poststructural turn in continental philosophy. Nevertheless, I consider that we have a lively native practice of critical cultural theory whose own institutional contexts cannot be read from afar—anymore than we can assume the current continental passion for reconstructing a tradition we never possessed. The cost of continental imitation, admittedly, is that we might lose those flash points at which a local institution demands its own revisioning of the general and particular understanding of the cultural history in which we are immersed. In this sense, then, the following chapters are defiantly "American" essays. This is because America is at once the homeland and the wasteland in which so many of them were conceived and because America remains the problematic place where what is learned from them is still teachable.

CAVE CULTURE AND THE NEED FOR TELE-VISION

All art nowadays claims to have no story line. Artists claim either to be absent from their work or else they claim that the work has eluded them— and is "unfinished." No self-respecting minor artist would betray either him/herself or the art work by giving it a title. The loss of stories is now the official line in all so-called postmodern culture. This is due in part to the seemingly irremediable non-sense of the affluent section of the capitalist world and for the rest to the death or disintegration of the great story of socialism. Western artists and intellectuals take it for granted that the story lines of religion are the opium of the masses from whom, however, they differ only by enjoying a distinctive narcotic style. It may therefore be an entirely superfluous gesture to address to any possible reader of these chapters a note on how they might be read. For this would be to reclaim intention and authority. But we do so. After all, to recognize that our thoughts are not entirely our own, or that we have not reached them by any royal road, does not dispossess us. Rather, this is the very circumstance of our perception, vision, and intelligence, as I have shown elsewhere.[6] No author proceeds without a possible reader in mind, if only because the author is necessarily a reader most of the time. Therefore what an author must keep in mind is that any possible reader might "re-write" what is written here—and that such an effect is to be hoped for and, indeed, invited in the present remarks. It is for this reason that I consider the introduction and the conclusion to this volume to function neither as respective beginning and end but rather as rival essays that threaten the very order of the intervening chapters and divisions of *Plato's Cave*. Their function, then, is to recast the topics I have settled upon just as I had myself to shuffle them several times before settling for better or worse on the present "hand." I believe this "play" in the text is there not because it lacks a strong logic—it could acquire that by reducing its arguments to the conventional logic of texts upon communications, media technology, or political economy, and embrace an even stricter logic if it aspired to be an exercise governed by the norms of narrowly professional publication which I have explored on another occasion.[7]

The intellectual force of the chapters in *Plato's Cave* derives from its determination to analyze matters that are "already known" because they have been "seen" a million times. It deliberately courts superfluousness by talking to a medium that demands not to be spoken to while it is speaking—in the-name-of-the-TV—a law that any child will invoke against its stuttering parents. Thus *Plato's Cave* is not about TV, or advertising, or mass culture—nor is it about government, economy, law, and the family—anymore than it is about feminism, Marxism, nationalism, tourism, and AIDS. It is, however, deeply concerned with these matters as sites where the life-world is colonized and conscripted into a larger system of controls that derives from no single center but is articulated through the therapeutic state, the corporate agenda, biotechnology, and a number of specular ideologies through which we communicate our desire to transfigure our carnal condition.

I believe that I am no different from any possible reader of *Plato's Cave* inasmuch as a large part of my life has been given up to reading a forest of newspapers, books, and journals when not spending light years watching TV and movies—not to mention a certain amount of immobile art inside and outside of museums and galleries. I may be peculiar only insofar as I am unable to "write off" these experiences as mere entertainment or as a considerable pleasure that need not be underwritten with the guilt of analytic interpretation. Admittedly, there is a considerable temptation to let much of the media go by rather than to "bore" those who celebrate it as a continuous rocky horror show. Indeed, postmodern critical theory might well convince one of the wisdom of sitting among the couch potatoes rather than slinking away to the study to reduce things to the lost arts of print. Alternatively, one can allow oneself to be seduced by the media fiction that one is learning something about the events and issues brought to our "attention" in the sitting rooms, bars, and lounges of the nation. This is an attractive myth for most people, and even quite seductive to extraordinarily literate persons, despite their awareness that "knowledge" is not easily won from an armchair or a bar stool nor while eating and drinking—and certainly not under the bombardment of commercials that celebrate every conceivable form of near-idiocy of which we are capable. And yet we are condemned to try to learn something from our media since almost everything we do is specularized in them.

All our appearances seem destined to begin and to end in appearances. This is the dilemma to which *Plato's Cave* is addressed. If we look behind our images we may find no solid reality. Yet, if we surrender ourselves to the "hyper-reality" of appearances, we shall lose the ability to discriminate light from darkness and thereby lose even our own shadow. We therefore desperately need some sort of *tele-vision* that is not a way of "seeing through" the media because of a refusal to be caught looking at them, but a way of seeing further, of seeing longer, and of seeing more steadily the risks we engage if ever we subordinate our intelligence to sensory modes of sight and sound whose own intelligibility has been colonized by agendas that waste the body politic. In this sense, "television" can only be nurtured provided that we consider all culture to be a *political drama* in which we have much to lose unless we are vigilant.

Since we are all affected by the media, they must be considered a political institution. For this reason, I have tried to formulate some general hypotheses regarding their function without, however, reducing their operation to a simple propagandist or conspiratorial agency. On the listener, viewer, reader's side, the political relevance of the media requires us to foster *tele-vision* by remembering that we are citizens as well as consumers, that we are concerned with our ecological and urban environment, with our families, our health and education, and with a legacy we have received from past generations that our own future has no right to destroy. We are concerned with exploitation, racism, and sexism and we still have some hope that all human beings will one day enjoy lives that are not impaired by the inequality bred by indifference to truth and justice. It is a conceit of the media that they too serve these ideals by reporting upon them, by documenting their failures, and by celebrating their successes—yet with a keen eye to entertainment and profitability. It would be unrealistic to disavow the complexity of the agenda that rules the communications industry. It would be just as foolish to overlook the fact that the complexity of its agenda requires of its audience a firm resolve to hold on to its own anchorage in those institutions that foster our vision of a civil and democratic society as the medium of all our messages.[8]

NIGHTLY NEWS: OILING THE WAR MACHINE

Americans have so reduced their imagination that they can conceive of no other way of life than their own. Moreover, their way of life in fact renders them incapable of living, thinking, or feeling otherwise than they do. All peoples have begun in this primitive state. Only Americans have managed to remain in it. They owe this achievement at once to the brevity of their history and to the inanity of their politics. American history has been short, nasty, and brutish, borne by a militant capitalism which emancipated the Southern agrarian slaves in order to complete the enslavement of the American consumer and the *petit embourgeoisement* of its workers. American politics have no basis in American society other than the religious hope of the poor that they may be included in the American Dream. The function of American politics is at once to awaken this dream and to disappoint its dreamers by appointing the elected rather than the elect. The equality of Americans consists in their tolerance of the ways they have of deceiving one another, impoverishing one another, and of murdering one another. The birthright of Americans is given with their congenital blindness to America.

Thus, strictly speaking, *Americans have no foreign policy other than a psychotic fear of not being loved as Americans.* This fear populates their world with aliens, communists, terrorists, and savages. America's weak military record requires that its wars—which cannot be lost because they are rarely official—be rewon by Hollywood's collection of soldiers, spies, and cops—with the useful conscription of "innocent" tourists as American hostages or, rather as *hostages to America.* Indeed, America's most official war must be considered to be its own Civil War. It is a war that both sides can claim to have won or lost depending upon one's point of view, rather like the so-called World Series in which only Americans play ball—however black. The Civil War continues today, of course, but it is fragmented into urban, environmental, crime, and drug problems. If anything in this seems cynical or unfair it is because the media have failed us all by reporting the news from "America." Part of the responsibility for the state of "America" must also lie with our schools, universities, newspapers, and magazines which have delivered up a totally untutored public mind to "MEDIAMERICA." Moreover, since our schools and universities and the publishing industry which services them are committed to the production of a docile citizenry to whom they appeal as a clientele, very little can be hoped from their side. This becomes clear every time America declares a war upon its own ignorance, poverty, and crime. America cannot fight "America" because the initial puritan excitement raised by such challenges soon becomes tainted with "Anti-American" prejudice. *No one in "America" loves an anti-American.*

The "news" would not, of course, achieve its specular tasks without the intermediary of advertising. The *commercial break* is in fact so crucial to the sense of America's news that we might regard the commercial as the reality that the news reports to be in jeopardy. There is nothing conspiratorial about this task—otherwise it would be well beyond the practical intelligence of the news agencies which interpret our society to itself by means of itself. Here, as usual, the news "shows" profit from the ideological capacity of daytime Americans to respond to such injunctions as, "Just Do It." There is, after all, no advice behind this recommendation, nor any commandment. It renders the consumer sovereign and spontaneous in a realm—footwear— where nothing is at stake, and at the same time it indulges a population that is going nowhere by convincing it that it is taking the first step by itself. The fetish of footwear in North America, like the proliferation of vans, jeeps, and off-road bikes, merely expresses a failure of nerve in the USAA (Automobile America). By the same token, anything that promises to get America "on the road" will indeed "mobilize" the American public—even if it is along the road to war. This is because war uses automobiles, trucks, planes, oil, and footwear and lots of American boys willing to defend the American home which in turn houses the fetish commodities of the American Way of Life, leaving aside that it is mainly blacks and poor whites who do the slogging.

American "wars" are essential to America's mood. This makes it hard to grasp that the first shots in the confrontation with Iraq over its conquest of Kuwait were fired by U.S. oil corporations against the American consumer.[9] The public could protest the increase in prices at the pump—but at the risk of Anti-Americanism. The use of news media to sentimentalize President Bush's slow dash to the Middle East by showing separated families guaranteed a weak opposition by critical Americans. Despite the extraordinary expenditures on oil to equip the ships and planes—not to mention the President's own faulty gas guzzling holiday craft, Fidelity!—the oil hungry American public attacked by its own "Eastern" corporations could hardly complain at the President's pact with the corporate/military complex to defend them against savage deprivation by the "Middle East." Throughout the so-called "oil crisis" in the fall of 1990, the "newsmedia" slavishly worked at the fiction of danger while the business pages of the newspaper belied the crisis of scarcity.

The military vacuum created by the East-West detente was immediately filled by American excursions into Panama and Saudi Arabia. The Panamian jaunt, at least, established that rock music is a military weapon as effective as chemical warfare! Iraq merely proved that American boys will go fishing even at the worst times, and that Canada's Prime Minister will do anything to jump on board with the United States, even while his own backyard is on fire! Curiously enough, Canada's backyard crisis with the Mohawk Indians—precipitated by the municipal proposal to extend a golf course at Oka on land containing a burial ground—presented the final stage in the specularization of government violence. Day after day, the newsmedia stood beside and *in between* face-to-face contacts between the Canadian Army and the Indian Warriors—their cameras creating a truly "pornographic" record of the encounters between two "mother-fuckers," staring each other down.

Rarely has the self-conscription of the "newsmedia" been so evident as in the "crisis" over—what? Kuwait, Iraq, the Middle East, Israel, Palestine, or the United States itself? Night after night, day in and day out, belligerent anchor men (including Barbara Walters herself) and supine "experts" adopted the "American" stance even though Congress was not even in session when Bush fired off his navy. At the same time, the degree of conviction in all of this was carefully monitored by the stock market, give or take a little for its own mood swings.

But all the while corporate interests in the movement of men and material—threatened with obsolescence by the end of the Cold War—were equally served whether or not a shot was fired since Bush's spectacle had already run up a bill of $23 billion in the first six weeks or so of the crisis.[10] There are, of course, some jobs involved in this, but the larger public does not hold shares in the "national debt," however much it suffers from it repayment. Worst of all, the world now faces the prospect of the American army playing *Globo-cop* in countries it no more understands than it does the jungle. We can therefore expect more and more tourists and "innocent" citizens abroad (women and children, the sick) to be used as pawns in the game. As a matter of fact, it is essential that there be hostages in this new global game since they protect the U.S. military against failure while they serve to sentimentalize and democratize U.S. militancy. Thus, if "even one" American life is saved or lost in a game created by Globo-cop politics, the price is right. Meanwhile, in America's own civil war with itself, thousands of Americans die from crime, poverty, and diseases against which America offers itself no protection. On the world level, however, it will be necessary to "beef up"

UN-cop as a side-kick to *Globo-cop*, just as white police forces conscript ethnics and women to implement a law that isn't always on their side.

The only reason that a sane person does not despair over these things is that America is not entirely ruled by "America." Indeed, to understand America we must create the fiction of "America" in the hope that it will sufficiently irritate our more intelligent and ethical self to work against the ways we have of impeding ourselves with the trash and nonsense that veil the corruption and exploitation which a few exercise against the many. This is the eye of the camera. It is the political blind spot that disables our intelligence and our morals while expecting the rest of the world not to see us any worse than we see ourselves. Yet we open the millennium with a psychotic renewal of our own star wars!

NOTES

1. *The Republic of Plato*. Translated with Introduction and Notes by Francis Macdonald Cornford (Oxford: Clarendon Press, 1941), Book VII, 514a-515c.

2. Allan Bloom, *The Closing of the American Mind: How Higher Education Has Failed Democracy and Impoverished The Soul's of Today's Students* (New York: Simon and Schuster, 1987).

3. Martha Nussbaum. "Undemocratic Vistas." *The New York Review of Books*, XXXIV (November 17, 1987): 20-26.

4. *The Republic of Plato*, translated with notes and an interpretive essay by Allan Bloom (New York: Basic Books, 1968), pp. 403-407.

5. Gordon S. Wood, "The Fundamentalists and the Constitution," *The New York Review of Books*, XXXV (February 18, 1988):33-40.

6. John O'Neill, *The Communicative Body: Studies in Communicative Philosophy, Politics and Sociology* (Evanston, IL: Northwestern University Press, 1989).

7. John O'Neill, *Critical Conventions: Interpretations in the Literary Arts and Sciences* (Norman: University of Oklahoma Press, 1992).

8. John O'Neill, *The Poverty of Postmodernism* (London: Routledge, 1994).

9. Peter G. Peterson, "The Price of Gluttony," *The New York Review of Books*, XXXVI (September 27, 1990): 38-39.

10. Jean Baudrillard, *The Gulf War Did Not Take Place* (Sydney: Power Publications, 1995).

CHAPTER ONE

Looking into the Media:
The Technology of Revelation
and Subversion

It is so difficult to think about the media that after some reflection we are likely to conclude that our difficulty derives from the fact that the media thinks itself or, rather, that what is not thought and felt and said through the mass media—if not through the elite media of literature and the arts—is not worth thinking, feeling, or saying. This impasse reproduces our general passivity before television, newspapers, advertisements, films, and sporting spectacles. One is either a player, a committed commentator, or a fan—but hardly ever is a place kept for the contemplative mind. Not to know what is going on in the media is to be out of it. To claim to know more than what is going on in the media than the media allow for, however, is to be out of joint with the form and content of the media. Critics of the media are exiles, or else they are allowed to strut their brief moment among life's killjoys, as a reminder of those higher things for which we have neither the time nor the taste. TV or not TV, that is the question!

Critics, or plain interpreters of the media, must decide to be or not to be. Failing silence, they must gather their own act, acquire a self, and imagine a society within which to think, to feel, and to speak otherwise—for awhile at least. I shall try in this chapter to think together an otherwise infinite variety of media models, scenes, sporting, and meteorological events, along with the seemingly commercial values that are delivered daily into our homes without any of us quite knowing how we came to live within such an avalanche of images. Our concern, then, is with how the media of newsprint, film, music, and television convey to us life itself and not simply how to live, how to eat, how to dress, how to love or to suffer, how to injure or to kill. Our task is to interpret how these media convey to us a life that is no longer an alternative life, or an amusement, or a fiction. For to suppose that the media are merely adjuncts to our living is to suppose that the family, the school, the church, and the university are central to our living. It is to suppose that adults know who they are and that they know what they want; it is to suppose that parents know more than children; it is to believe that men and women understand the difference between social relations and sexual relations, the difference between love and cruelty, between commitment and contract. It is to suppose that we understand the difference between the state and society, between the home and the economy, between education and entertainment. It is to believe that men and women can tell the days apart and that they still hope for some difference between today and tomorrow, not because they wish to leave today behind, but because they wish it to be remembered with the rest of the past.

It is the daily business of the media to confirm these beliefs while destroying them. Thus our follies are never beneath reason and our reason is never more than a limited folly. Therein lies the field of phenomenological criticism, which accepts the convivial limits of reason while nevertheless claiming to be authentic. Such criticism has its counterpart in the celebration of everyday life begun by Montaigne, Proust, and Joyce, for example, inasmuch as the literary redemption of the profundity of ordinary living can be achieved only through sounding dreams and myth.[1] In other words, I have in mind in the following arguments a concern for critical understanding that does not exploit the differences between the way things are and the way they might be. Rather, I wish to leave open the possibility of their reversal through our care of what is sublime as well as what is desperate in our affairs and the times we are living through.

To make any beginning in thinking about the media, we need to consider technology as revelation, to regard our machines as the prime instruments of appearances, of dreams, of visions, and of truth in the assertions of skyscrapers, cathedrals, banks, theaters, automobiles, television, newsprint, rock music, refrigerators, and contraceptives that furnish everyday life with its average sense and nonsense. Therefore, it is ourselves we see in television and in print commercials because we express ourselves in these media only with greater exaggeration than in the rest of our living. To show what I have in mind, I propose to interpret several communicative artifacts—a paper mat, a potpourri of commercials, and some misreadings of rock, open theater, and urban graffiti. I shall try to find in these varied communicative media how it is we reveal and subvert the deepest, the most trivial, the most holy, and the most vulgar of our beliefs and values.

If Woodstock, as Abbie Hoffman has remarked, was an attempt to land man on earth, then that even larger celebration of America's birthday must be viewed as a more determined effort to land man right in America. We shall return to Woodstock. But let us begin with the Bicentennial as a communicative artifact in which the commonest things express to Americans who they are in what they eat and, fundamentally, in how they choose to eat.

THE MEDIA OF OUR DISCONTENTS

The Declaration of Independence provided Americans with a machinery of *divine discontent* without which we cannot understand American political economy:

> We hold these truths to be self-evident: that all men are created equal; that they are endowed by their creator with certain inalienable rights; that among these are life, liberty, and the pursuit of happiness; that to secure these rights, governments are instituted among men, deriving their just powers from the consent of the governed; that whenever any form of government becomes destructive of these ends, it is the right of the people to alter or to abolish it.[2]

I shall argue that *America is built upon a communicative machine known as the American Way of Life*. This machinery operates a process of *commodification*, which materializes the ideals in the Declaration of Independence in an endless flow of deals in the marketplace. Because of the massive rhetoric of the American Way of Life, there is no place for any other revolution than the first American Revolution. The result is that whereas in Europe everyone is revolutionary but few are subversive, in America no one is revolutionary, though many are subversive. This was first observed by Alexis de Tocqueville, and I believe his observations are as good today as they were over 100 years ago:

> I have often remarked, that theories which are of a revolutionary nature, since they cannot be put into practice without a complete and some- times a sudden change in the state of property and persons, are much

less favorably viewed in the United States than in the great monarchal countries of Europe; if some men profess them, the bulk of the people reject them with instinctive abhorrence. I do not hesitate to say that most of the maxims commonly called democratic in France would be proscribed by the democracy of the United States. This may easily be understood: *in America men have the opinions and passions of democracy; in Europe we have still the passions and opinions of revolution.*[3]

The truth of Tocqueville's remarks has been marvelously absorbed by the commercial media of the American corporate economy. I can only begin to explore this phenomenon. Consider Figure 1.1 (*A Declaration*). It is a paper place mat from a large chain restaurant, Howard Johnson's, a landmark all over America, famous for its 28 varieties of ice cream. Howard Johnson's is a sign among all the other road signs that show Americans where to gas up, feed, amuse, sleep, and toilet as they move along highways whose only respite from the speeds they afford is the signposts that offer to just as speedily service the weary traveler in order to return him to the highway. Howard Johnson's is a traveler's icon, the same everywhere one goes, offering the same variety of food in the same restaurant, welcoming the traveler free to roam in one of the world's most orderly traffic flows. America's Bicentennial celebration was not at all desecrated by Howard Johnson's use of the Declaration of Independence. On the contrary, the desecration of the American creed—its commodification—is essential to the daily practice of American democracy. As long as the rights to life, liberty, and the pursuit of happiness can be translated into the commodifications of ice cream, automobiles, travel, and smoking, enjoyed by a public willing to consume the

shoddy lookalikes of mass production in their homes, clothing, amusement, and education, the American dream is given an individual fit. It makes itself something that people can grasp in a tangible form. At the same time, everything Americans do grasp must also reflect their aspirations to higher things enjoyed in this world.

Ice cream is a perfect example of the artifactuality of the American dream. Ice cream is a hybrid of nature and industry. It fulfills a desire that by itself neither nature nor industry—or even our own mother—could satisfy, except for the know-how that brings them all together. It is a gift that recruits children forever into a world of endless variety from whose cares it will later free them along with all the other self-rewarding foods and drinks that make up popular eating habits. *Ice cream*, if not the ice cube, *is the totemic food of Americans*. It accomplishes the infantilization and permanent communion of Americans in the variety of choice and the happiness of a food that symbolizes the unity of nature and industry produced playfully as a bounty of the promised land. Americans easily identify their rights to life, liberty, and the pursuit of happiness with their quest for wholesome food available in endless variety and at very little cost. It is central to the American way of life that the quest for food not symbolize human suffering and exploitation. It is this that distinguishes them from foreigners and natives and thereby motivates American charity.

Consider Figure 1.2 (*The American Way with Food*). The text is remarkable for the ease with which it makes the choices of religion, occupa-

tion, and government resemble the choices consumers make in a supermarket. The identification of food choices with religious, economic, and political freedom in turn overdetermines the belief that the latter institutions are in a gross way the outcome of the same individual choice and caprice that go into buying cakes and cookies. The powers of church, state, and economy are denied any determining influence upon the sovereign consumer. Thus the harsh realities that govern the everyday obligations of law, labor, and belief are removed from consciousness in favor of a down-to-earth interpretation of the ideal of freedom as the American way with food. In exchange for his ignorance of the underlying realities of the American political economy, the consumer is made wise, pragmatic, and omnipotent in the exercise of trivial choices of breakfast and snack foods. The products themselves are made to appeal via the tropes of family, tradition, fun, health, and the rights of children. They make ordinary things and occasions into special and extraordinary experiences. They are made by a large corporation with a homey face, dedicated to the family and the kids, relieving mother from baking while instilling in her the feeling that she might have baked all the good things on the Pillsbury table.

The commodification of the American Revolution collapses the historical awareness of Americans, while at the same time reviving it in the commercial sagas of the wild West, antique hunting, and the retailing of the American heritage in volumes that decorate coffee tables or provide ammunition for the tireless projects of competitive schoolchildren. If the past is to survive, it must contribute to the future by seeming to be either an irresistible impulse to national progress or a collection of handy facts for children on the make in school. Thus, in the popular mind, American history owes far more to the movies than to the poor competition of its school book versions. In the first place, there was the promised land of the early settlers. Admittedly, the land had to be cleared of Indians. But even this work was democratized by the six-shooter and the rifle. Later, the tractor cleared the white farm families themselves from the land, as corporate and industrialized agriculture began to feed the cities and the factories of America. If there was a victim in the wars against the Indians or in the Civil War (it took the Vietnam War to give the blacks their roots in white America) it was the farmer whom America today, forgetful of the grapes of wrath, remembers only as a six-shooting cowboy—or as Bonnie and Clyde, shooting at a world of corporate finance and controls they cannot grasp. Indeed, the cowboy and the gangster—the poor boys of town and countryside—hold the American imagination precisely because they are innocent of the ways of corporate organization, yet are devoted to the same ideals of success—a big spread, a huge ranch, a big car, and a big cigar—that are packaged in less heroic and more routine fashion for the corporate worker and consumer.

If the huge fortunes and the corporate organization of country and western music[4] do not disturb the downtrodden and divinely rehabilitated images of its heroes, it is because *country and western music audiences are the loyalist branch of the American dream*. In the same vein, the success of the Mafia-inspired *Godfather* films is intelligible if we understand that they restore successful violence to the American family, reviving ethnic loyalty as the enemy of anonymously rational corporate organization. The formal similarity remains even in *Dog Day Afternoon*, which joyfully celebrated the bungling criminality of an Italian, Catholic homosexual at grips with the organized overkill of the police. The movies are an American artifact. They are as necessary to the communication of American political life as the hot dog, the automobile, cowboys and Indians, and cops and robbers are elemental to the community they portray.

The movies are a continuous American mass, celebrating the great theme of individualized success and failure that is also the daily bread of the American cartoon strip.[5] The movies are the depoliticized revelation of the American Revolution. However momentarily, they recruit everyone into the popcorn lumpen-proletariat that dreams of a sudden reevaluation of its original rights to participation in the American dream. The American passion for documenting failure, for celebrating the underdog as much as it celebrates success in business, sports, crime, and beauty contests, is a passion for documenting the Declaration of Independence. Moreover, it is a passion that makes politics fun—it allows the business of keeping an eye on the American charter to be carried on as an amusement activity, the pastime of paid-up members consuming the offerings of legitimate theater or journalism. For these reasons, the newspaper and the movies are perfectly integrated communicative mechanisms of depoliticized conscientiousness, inasmuch as they are consumed in private, familized settings, in free time, and thus removed from the political motivations of a real public.[6] The spectator who contemplates the saga of success and failure learns from failure that success is better. "Success," of course, is not the sheer matter of winning; it is the practice of freedom and abundance on individual terms. It is a practice that requires a world of opposition that contains only other individuals or the sheer otherness of nature (including primitives and colonials) red in tooth and claw. Thus the myth of success and failure is a *depoliticized myth*—the Revolution as a Way of Life—that hides the stratifying realities of class, racism, and corporate power that otherwise shape the American's exercise of his/her rights to life, liberty, and the pursuit of happiness.

Commercial Bodies

Marx argued that all production is social. I want to include in the notion of production not just the body's physical labor, but every experience, sensation, and activity of the body as a field of production and consumption. By this I mean that we must regard the *productive body*[7] as an extension of the economy and not simply as a factor of production, like labor. Like labor power, the productive body exists only in a market economy, which is capable of reifying its stress, relaxation, health, illness, beauty, spontaneity, and sexuality. The reification of the body within industrial and commercial sectors concerned with its production and consumption integrates and distributes the body throughout the economy. The productive body is not a factor of production in the way that Marx envisaged land, labor, and capital. The productive body is integrated into the division of labor, both internally, as through modern medicine, and externally, as through fashion and cosmetics. Thus the productive body is both an extension and an intensification of the space and activity of the modern economy. It is not simply that the economy expropriates the labor of the body, subjecting it to pain in its tasks and to an unsatisfactory standard of living in return for its wages. The most massive exploitation of the body occurs as the economy teaches us to disvalue our body in its natural state and only to revalue our body once it has been sold grace, spontaneity, vivaciousness, bounce, confidence, smoothness, and freshness. The media are a principal socializing agency in those *techniques of the body*[8] that display the cultural values of youth, aggression, mobility, and sociability. By the same token, the media are obliged to hide the ordinary condition of men, women, and children, except as a shock or removed misery. In all this, television unites the family the better to divide it. The more the modern family unit is geared to consumption, the more it needs to separate into wage earners who separate sexuality and reproduction. Thus the female body must be deromanticized and made the instrument only of rational, that is, contractual associations. Trusting to the pill, young women's bodies are made mobile for work, high-rise living, and adventure. The physical hazards, ranging from cancer to rape, are all part of this bodily complex that sings its appeal to the young self-possessed women of the cigarette and perfume world.

We shall not understand the spectacular functions of the communications industry unless we think of the hold it has upon us as literally a grip upon our bodies, turning the body into a theater of corporeal agony and ecstasy. We must realize what the media realize, namely, that the body is the theater of all our desires, of our salvation and destruction, of beauty and ugliness, of joy and pain. The body is the seat of science and of religion, of psychoanalysis and politics. It is above all an endless topic and resource for the communicative media:

The body, like the body politic, is a theatre; everything is symbolic, everything including the sexual act. The principal part is a public person taking the part of the community as a whole: *persona publica totius communitatis gerens vicem*. The function of the representative organ is to impersonate, incarnate, incorporate in his own body the body politic. Incorporation is the establishment of a theatre (public); the body of spectators depend on the performance for their existence as one body.[9]

The body is the theater of our social life. It projects the spectacle of our self-presentations to others as we would wish to be seen by them, as well as being the mirror in which we seek our most private self. The human body is the holy place of expression and concealment; it is the instrument of our transcendence and togetherness, as well as the seat of our withdrawal and sorrow. Our body is our circumstance and our fortune; it is given to us by others, and yet, like nothing else, it is our own. Our body is the seat of our most persistent needs, weighted by the legacy of our ancestors, and yet lightly at the whim of our most fleeting desires, courting the stillborn future of fashion and obsolescence. For the most part we encounter the body when it is too late—overweight, in love, pregnant, or with a broken leg. We experience our body as something uncanny, something suddenly reversible—the cynosure of eyes, sweeping down a staircase to be met by admirers, the source of our shame and embarrassment, should we trip and fall, more like an apple than a star.

As we have learned from countless commercials, the spectacle of the body occupies much of our daily efforts. Beginning with the call to rise and shine, we proceed to wash, polish, scrape, and spray until we are satisfied that we are cheerful-looking, businesslike, alluring, handsome, and even sexy for the next 24 hours. Each of us can fantasize his or her own approach to these body rituals. We become Roman empresses when sunk in a bubble bath, or for women who find this too decadent, there is the virgin's struggle with pimples, stringy hair, mouth and body odors, all of which have to overcome in the half hour before the arrival of the knight in shining armor, Wilkinson's sword, and high karate. For older men, there is the massage parlor, provided it is open to the rush of unmannerly children berserk with their latest cavity count. For the dirty Irish, Erin grows green soap! The commercials have also taught us what as mechanical and urban men we need to know of nature. It is usual now to speak of the most recent stage of industrial society as one with special features—summarized in the experience of private affluence and public squalor.[10] No contrast is better suited to commercial scenarios.

Affluent society is by and large an indoor society. We spend our time inside factories, offices, clubs, automobiles, and houses. These interiors are also machine settings in which most of what we do is done for us. The machines have the power and we have the flab. Precisely because affluent society is an indoor society, we observe a massive exodus from its great dormitories to the countryside, lakes, beaches, and mountains. City dwellers are sports and nature lovers, as well as hi-fi bugs, bookworms, and movie-goers. How are we to understand this? A bewildering variety of vocabular-ies motivate the quest for nature, sport, and leisure, ranging from positivism and romanticism to militarism and existentialism. People find God and country in the great forests, in the surf, and on snowmobiles. They also find in nature a larger arena for the national pastime of littering, which has its great celebration in the forest fire. Other favorite activities in the great out-doors include the pollution of rivers and lakes with the noise of motor boats, as well as other forms of human dirt. The great outdoors is classically a man's world, or at any rate it belongs to our John Waynes as well as to our Tom Sawyers. It is a white man's world, though Negroes and Indians may tag along or serve as bait. Bears, golfballs, and fish serve as pretexts for men to go off together for days on end, under the sponsorship of their favorite brewery, to celebrate the great male bond to which the human race owes its survival. Women, though better left at home or parked in the coun-try clubs, function best in the outdoors as cooks or as symbols of the renew-al of the human race. But they do not make friends; this is a man's business. Friendship is found in fieldhouses, locker rooms, sweat and sneakers, pinch-es and pats with the boys. "Friendship is the clean-cut, competitive horsing around of regular guys, for Christ's sake!" Girls and wives can never under-stand this, and so it is better for them to visit mother, or go shopping, take evening courses, and leave the boys to their night out. In this way, we learn much of the rhetoric of our daily morality, and not only of our vice and vio-lence. When we do let the girls in on the game, it is only because corporate culture finds such accommodation opens up new lines of profit. Once in, women play the same game.

We should not overlook that the media have more serious concerns. Today, more than ever, the tragic spectacle of the body thrusts itself upon us in the media reports on the victims of war and civil violence, in the images of helpless victims of famine and other disasters. It is above all in the ago-nies of sport that the human body provides us with a morality play suited to the rise and fall of industrial man. In the early industrial period, the body spectacles commonly associated with sport and leisure were the privilege of the surviving landed aristocracy and the new middle classes. What could be learned on the playing fields of Eton may hardly have seemed civilized to the French, but it proved itself on the battlefields of India, for all its failures in the bedrooms of London and Paris. The gradual extension of sports to the lower classes, by and large such lower-class sports as football and visits to the public baths, depended upon the introduction of the five-and-a half-day working week—the remaining half day being given to Saturday football, beer, and dancing, with the Puritan provision of the whole Sunday for repentance. Today, commercially punctuated sports spectacles compete with the private agonies of sex and drugs scored upon the body. For technological man, the places of victory and defeat are by and large unspectacular— witness the failure of the greatest show on earth once the Cape Kennedy stagings became regular productions. By contrast, modern professional sports, and I would include here the Hollywood musical and local strip joint, have become the secular rites of industrial society.

Sports spectaculars are games of position and points, in which violence and efficiency are sublimated into a statistical code of victory and defeat, which in turn animates a largely passive audience. *Sports statistics are the history, law, and science of the average man*—the proof that he does not live by hot dogs and popcorn alone. Modern games are the soul of objectivity: they culminate in the passionless play of the moon game and the armaments race. Games, then, are the highest moments in the media world; they are indistinguishably communication, consummation, and consecration. Sports spectaculars are essentially ritual performances because the very nature of the escape they involve commits the viewer to the formal, professional, and technical-legal efficiency involved in their production. The same is true, I believe, although I shall not develop the argument here, of the new forms of sexual play and pornography. In all these activities, the body strives to become the machine that threatens to replace it in everyday life. In endless replays of the aggressive and competitive game of industrial society, sports spectaculars encourage the fantasy that the virtues and vices of industrial society can be mastered as character and guts. Women in sport play the same game.

If outdoor sports have any serious rival, it is from the indoor sport of weather watching. Nothing is better suited to people who spend their days moving from one container to another—from house, to car, to bus, to office, to bus, to car, to house—than wondering what it feels like in between. *The less we experience the weather the more we are concerned with it.* It is the perfect obsession. When most people lived on farms they could tell the weather for themselves and with as much predictability as can today's meteorologists. Indeed, prior to television weather forecasts, even urban people could tell the weather. As it is, the weather man, or weather woman, sits indoors and, with the aid of complex technical equipment, maps out the ups and downs, highs and lows of the atmosphere in a nightly school lesson, repeated for the last time of a hundred times a day. But what is the lesson? Do we still yearn for the outdoors, for the body's rhythmic ties to nature and the cosmos? Or is nature the last obstinate vagary in our lives, the hated and admired vehicle of the uncontrollable and unpredictable? If it is, then we can understand our pleasure in the meteorologist's misfortunes. His failures, like her irrelevance, are witness to our ambivalent love of science and technology, to our guilty conspiracy against nature. Having constructed a society in which we have nothing in common, the weather remains the last of all our commonplaces. As such, it is stronger than religion and more enduring than the technologies whereby we struggle to harness its forces.

TOWARD A THEORY OF COMMUNICATIVE SUBVERSION

Rather than sharing, Americans have preferred to interpret their revolution to themselves in terms of raising the standard of living[11] for all who can successfully compete in the massive organizations of industry, agriculture, and government that sing of the American Way of Life dispensed through toothpaste, nonreturnable bottles, automobiles, telephones, package tours, credit cards, and political conventions. The engine of the American Revolution has been the large corporation, now nearly global in reach and virtually sovereign in the integration of its investment, production, and consumption strategies. Whenever the soulful corporation appears unequal to the American dream, it can count upon the state administration of antidepressants to "revive the economy" and to get the American people moving forward. Although this economy is the young, the old, and the blacks—all of whom must compete with a highly skilled or programmed technology and its continuous "sophistication"—there has been little determination to alter the American concept of revolution, or to redefine it in terms of a radical change in the political economy of the rights to life, liberty, and the pursuit of happiness, except for a little recycling.

It is in the city, which we must regard as another of our machines for living, where the commodification of the American Revolution is most clearly articulated and where the sanctions of individualized success and failure are also most harsh. Urban life demands from many citizens a daily countercultural response. It requires a subversive practice of institutional effacing.[12] By this, I refer to a civic practice whereby the official and ostensible purposes of institutions, places, objects, and processes are substituted or effaced in favor of the proceedings of those who come to use them day by day or only briefly and, as it were, in flight. Thus, large institutions such as hospitals, airports, and universities, places such as hotel foyers, walls, and sidewalks, and things such as cartons and tin cans are all subject to native use, or *bricolage*, that effaces their ostensible meanings in favor of secondary creations and employments. The subversion of the official and planned uses of institutions, places, and things includes the practices known to them of schoolchildren, auto buffs, bums, welfare hustlers, happenings, and subway graffiti. Effacing is especially an urban art of transforming the iconology of industrial and commercial life to meet the real needs and experience of the city's poor, its aged, and its young people—who so often compete unsuccessfully with such middle-class practices of effacement as the gentrification of inner-city neighborhoods, which import the nostalgia of poverty while expelling the poor.

The urban environment, and in particular the ghetto, is the prime target of effacement. It is saturated with signs that fixate and stereotype ready-made experiences, objects, and encounters that stylize the bodies, dress, and vehicular movements of city people.[13] The aesthetics of the city are a flashboard of the exchange values and the icons of commodification that celebrate the commercial life of its denizens. By the same token, the practical aesthetics of the city are noisy, ugly, and hostile to those who are incompetent with the city's official uses and occasions. Everything that is contradictory and incoherent in the material basis of the political economy is reflected in the neonized iconography of the urban environment, in its wealth and poverty, its comforts and dangers, its crime, its sophistication, and its vulgarity. The city is hard on those who are not making it in the city; it crushes and silences them unless they are able to subvert it by creating their own style and unforeseen ways of holding out, getting by, and hitting back. Thus the city is open to the endless necessary profanations and effacements whereby persons otherwise excluded from its ostensible activities make a place for themselves, cut a figure, and hold out where they would otherwise seem to be submerged. It thereby furnishes a prime genre of television series, from *Kojak* to *Police Woman* to *Miami Vice,* and more recently *Law and Order* and *NYPD Blue* that serve to spectacularize the drama of law and order as an urban morality play. (We return to this theme in Chapter 8.)

Walter Ong has argued for a certain parallelism between the sequences of communication processes and the Freudian stages of psychosexual development.[14] There are, of course, huge problems in such comparisons, not to speak of the initial abstraction of such a schema. But what is of value in these constructions is the notion that the human sensorium is a structurally and historically produced producer of its own acoustic, visual, tactile, libidinal, and social environments.[15] Thus it happens that our machines can alter the ratio between the human senses as well as between society and nature. In what I have called *gay technology* in an attempt to interpret the technological display accompanying rock music, I believe we may discern a variety of social forms that return the industrial and political "machine" to the rhythms of life and the body politic. Along these lines, I think Woodstock was an attempt to give birth to a new American nation (the Beatles had the same role in the UK) through a profound subversion of old-order machine culture, technological decorum, and occupational identity. Rock music, its artists, and its audiences dissolve the sterility of technology into the convulsive improvisations of violence, love, care, and community. Rock like pop expresses the joyful embrace of life and technology that is the driving vision of the modern world, as well as its own nightmare. The spectacle of rock reveals the world as desire, the body as environment, caught between order and chaos, invoking community, flirting with self-destruction and infantile disorders. But the conventional technological congregation demands decorum in the presence of machines, engines, typewriters, microphones, and television sets. For this reason rock is especially revolting. Its standards of technological decorum are flirtatious, cajoling, argumentative, burlesque, and destructive. Yet it is now clear that rock is as much part of the global money machine as any other business it appears to reject.

In modern society the ratio of self and expressive space is so radically diminished that we experience our bodies as shrunken landscapes. Thus, in our literature the typical setting of the modern self is the underground, the tiny room, the prison, the asylum, and the concentration camp. These are the scenarios of modern sociology and of modern theater. We see the self trapped in primitive settings that force it to account for itself as an object, obsessed with the degrees and forms of distance, bewildered by the fear of others who anticipate the slightest deviation in standards of spontaneity, cleanliness, generosity, and courage. It is no wonder, therefore, that these underlying anxieties furnish the materials of so many of our body commercials, as I have tried to show.

The cultural and communicative forces for depoliticized integration and legitimation which I have described as the commodification of the American Revolution make it difficult to discern any long-term trends in the communicative subversion of the American corporate economy. Meantime, it is clear that the Marxist theory of revolution is inadequate because it continues to locate the impulse to resistance and change in the exploitation of productive labor, whereas it is the generalized code or semiotic[17] of the administrative rationalization of economy, society, and polity that must be identified as the point of attack and subversion. An adequate theory of subversion must, therefore, be a theory of the communicative processes of political economy that now mediate production and consumption, language, sexuality, urbanism, feminism, suburbanism, nature, culture, youth, and old age. In short, the theory of communicative subversion is a theory of the political economy of the body. It supersedes the Marxist theory of revolution because it generalizes the semiotic of exchange values to every bodily and mental conduct beyond the simple productive labor required to mobilize commitment to a rationally administered economy. It thereby opens up the field of inquiry begun here.

WILD SOCIOLOGY

The question of the relation between cultural subversion and political revolution is one that appears to divide us into realists and dreamers. Here, at least, appearances are not deceptive. The options of political theory are indeed orders of character. Cultural subversion is avowedly utopian. It is inventive of character and society. In this, however, it is rejected as inadequate knowledge, evidently innocent of sociology, economics, and politics. What is at stake is the unborn sociology of a society reflexively aware of its notions of order and character. In making this issue plain, cultural subversion toys with the dissolution of professional social science and its expert/lay organization of knowledge. It reveals the distance between sociologists and a sociology aware of itself as work with people. Such a sociology dreams of an end to the hierarchy of knowledge suspended in a genuine collectivity of social work. It is a wild sociology—not in the sense that it is prehistorical sociology—but precisely because, within the very history which wild sociology presupposes, it dares to be utopian. The ignorance that determines professional sociology, on the other hand, is precisely its unhistorical knowledge of the present—modified, to be sure, by its construction of history as the past, but never illuminated by a projection of history as utopia. Whereas establishment sociology is concerned with the administration of existing social order, wild sociology is free to project scenarios of alternative orders. This is what is at issue in the crisis of Western social science and the society that it reflects. It is for these reasons that individual awareness of the quality of everyday life—its objects, language, space, time, and needs—erupts into the meaninglessness of the corporate agenda. The politics of experience represents the insurgency of human values at the low points of everyday life in the urban industrial world, at what Henri Lefebvre calls the "zero point" of social experience where a kind of irrational asceticism is discernible under the apparent affluence and rationality that dominate our lives, but never so completely as to make countercultural response impossible.[18]

This is not to argue that cultural subversion is the only way of revolution today. However, that is not because there is some intrinsically "political" strategy of revolution. I mean rather to remark upon the poverty of our culture as a resource for revolutionary transformation. This is to weigh its failure, its class bias, its fragmentation and frivolity. It is to feel the nervelessness of a culture that is the property of experts and entertainers, prostituted and destructive of the very style of life that is the underlying promise of all culture. At the same time, it is to touch at this very zero point of culture its promise of the transfiguration of everyday life, not as a canvas to be wiped clean but as the natural light of humanity.

NOTES

1. John O'Neill, *The Communicative Body: Studies in Communicatiue Philosophy, Politics and Sociology* (Evanston, IL: Northwestern University Press, 1989).

2. *Time*, Special 1776 Issue (1976), p. 8.

3. Alexis de Tocqueville, *Democracy in America* (New York: Vintage Books, 1954), 2:270. My emphasis.

4. "Country Music, Songs of Love, Loyalty and Doubt," *Time* (May 6, 1974), pp. 51-55.

5. Arthur Asa Berger, *The Comic Stripped American* (Baltimore: Penguin Books, 1974).

6. How "prime time" is divided into a complementary vision of public decisiveness in the crime series and private fumbling in the sitcom is nicely explored by Joyce Nelson, *The Perfect Machine: TV in the Nuclear Age* (Toronto: Between The Lines, 1987), Chapter 3, "Atomic Fictions."

7. John O'Neill, *Five Bodies: The Human Shape of Modern Society* (Ithaca: Cornell University Press, 1985).

8. Marcel Mauss, "Techniques of the Body," *Economy and Society* 2:1 (1973):70-88.

9. Norman O. Brown, *Love's Body* (New York: Vintage Books, 1966), p. 137.

10. See Chapter Two.

11. David M. Potter, *People of Plenty: Economic Abundance and the American Character* (Chicago: University of Chicago Press, 1955), p. 126.

12. I have derived the notion of effacement from the works of Marcel Duchamp and Rene Magritte. In 1913, Duchamp exhibited a "ready-made" *Bicycle Wheel* and, in 1919, *L.H.O.O.Q.*, a bearded *Mona Lisa*, a modified ready made. See Arturo Schwarz, "Contributions to a Poetic of the Ready Made," in *Marcel Duchamp: Ready-Mades, etc.* (1913-1964), Walter Hopps, Ulf Linde, and Arturo Schwarz, eds. (Paris: Le Terrain Vague, 1964), pp. 13-38. Magritte's work of displacing (*dépaysement*) commonplace objects was influenced by Giorgio de Chirico's *The Song of Love* (1914) and Max Ernst's work in *frottage* and *collage*. See James Thrall Soby, *René Magritte* (New York: The Museum of Modern Art, 1965), p. 8. It is difficult to single out works of Magritte. I mention only *The Key to Dreams* (1930) and *Personal Values* (1952).

13. John O'Neill, "Lecture visuelle de l'espace urbain," in *Colloque d'esthetique appliqué à la création du paysage urbain*, Michel Conan, ed. (Paris: Copedith, 1975), pp. 235-247; John Berger, *Ways of Seeing* (Harmondsworth: Penguin Books, 1974).

14. Walter J. Ong, *The Presence of the Word: Some Prolegomena for Cultural and Religious History* (New York: Simon & Schuster, 1970), pp. 92-110.

15. John O'Neill, "On the History of the Human Senses in Vico and Marx," *Social Research* 38 (1976):837-844.

16.   John O'Neill, "Gay Technology and the Body Politic," in *The Body as a Medium of Expression*, ed. Jonathan Benthall and Ted Polhemus (London: Allen Lane, 1975), pp. 291-302.

17.   Jeremy J. Shapiro, "One Dimensionality: The Universal Semiotic of Technological Experience," in *Critical Interruptions: New Left Perspectives on Herbert Marcuse*, ed. Paul Breines (New York: Herder and Herder, 1970), pp. 136-186; Jean Baudrillard, *La société de consommation, ses mythes, ses structures* (Paris: Gaillimand, 1978).

18.   Henri Lefebvre, *Everyday Life in the Modern World*, trans. Sacha Rabinovitch (London: Penguin Press, 1971), p. 185.

CHAPTER TWO

The Corporate Mediation of Private Lives and Public Apathy

Whereas the age of communication promises to be an age of democracy, the truth is that the *media titillate us but do not educate us with genuine politics*. This is because our political imagination is shackled by the corporate agenda that shapes modern political economy. The traditional antitheses of individual and state, state and society, public and private rights, conflict and order no longer serve to orient men's private lives toward their political contexts. *Modern society is increasingly consensual and apolitical*; it generates a comfortable reality which tempts us to identify the rationality of its industrial metabolism with the whole of rationality and thus to disengage ourselves from the critical tasks of reason. The tendency to identify technological rationality with social rationality is the major threat to the survival of the political imagination.[1] It underlies the liberal abdication of politics in favor of the market economy.

Liberal political economy remains nerveless so long as it rests upon a concept of government which does not question the social distribution of resources between the public and private sectors of the economy. No modern government can retain power which fails to control industrial technology and the power of large corporations to shape the national ecology and psychic economy of individuals, which they are able to achieve in part through the media as the mirror of the world and lifestyles that the corporate agenda requires us to observe. The corporate economy stands between the state and the individual. Its power to determine the lifestyle of modern society must be recognized as a major subject of political economy. The critique of the forces working to produce what Herbert Marcuse has called one-dimensional society must avoid the elitist fiction that mass society is the cause of our political troubles as well as the liberal illusion that pluralistic countervailing power is the only viable formula for political conduct.[2] At the same time, the basic organizational form of modern industrial society is so closely tied to such a small number of corporate and bureaucratic structures that the ideas of pluralism can hardly be said to exercise a qualitative effect upon the system:

> In a specific sense advanced industrial culture is more ideological than its predecessor, inasmuch as today the ideology is in the process of production itself. In a provocative form, this proposition reveals the political aspects of the prevailing technological rationality. The productive apparatus and the goods and services which it produces "sell" or impose the social system as a whole. . . . The products indoctrinate and manipulate; they promote a false consciousness which is immune against its falsehood. And as these beneficial products become available to more and more individuals in more social classes, the indoctrination they carry ceases to be publicity; it becomes a way of life. It is a good way of life—much better than before—and as a good way of life, it militates against qualitative change. Thus emerges a pattern of one-dimensional thought and behaviour in which ideas, aspirations, and objectives, that, by their content, transcend the established universe of discourse and action are either repelled or reduced to terms of this universe. They are redefined by the rationality of the given system and of its quantitative extension.[3]

One-dimensional society is characterized by a systematic linkage between the subordination of public space to private space through the agency of the corporate economy and an ideological privatization of individual sensibilities which reinforces corporate control over the allocation of social resources and energies. One-dimensional society has its roots in the liberal concept of society as a field in which the private pursuit of economic interests produces public benefits without political intervention. The emergence of a "social universe," which is, strictly speaking, neither public nor private, is a modern phenomenon that arises from the public significance accorded to the business of making a living and has no counterpart in the ancient world. It is a phenomenon that has forced upon us the hybrid term "political economy" and with it the challenge to rethink the relation between the public and private domains in modern industrial society. *Because television amplifies the social universe, it seriously reduces our political vision.*

PUBLIC AND PRIVATE SPACE

In the Greco-Roman world, the boundary between the public and private realms was clear and men were conscious of the threshold between public and private life. Although the ancient city-state grew at the expense of the family household and kinship group, the boundary between the public and private realms was never erased. Indeed, the definition of the public realm as an area of freedom and equality presupposed the recognition of "necessity" in the household economy.[4] The needs of maintenance and reproduction defined the social nature of man and the family, and the sexual and social division of labor between man and woman, master and slave.

In the modern period this ancient boundary between public and private realms was dissolved with the emergence of "society" and the liberal concept of minigovernment. A whole new world—the social universe—emerged between public and private life. The public significance of the social universe has its roots in the subjectivization of private property and the subordination of government to a minimal agenda in the social equilibration of individual and public interests. The 17th century reduced the political domain to the narrow limits of "government" in order to exploit the boundless domain of possessive individualism, which C. B. Macpherson has described as the central impediment of modern liberal-democratic ideology:

> Its possessive quality is found in its conception of the individual as essentially the proprietor of his own person or capacities, owing nothing to society for them. The individual was seen neither as a moral whole, nor as part of a larger social whole, but as an owner of himself. The relation of ownership, having become for more and more men the critically important relation determining their actual freedom and actual prospect of realizing full potentialities, was read back into the nature of the individual. . . . Society becomes a lot of free individuals related to each other as proprietors of their own capacities and of what they have acquired by their exercise. Society consists of relations of exchange between proprietors. Political society becomes a calculated device for the protection of this property and for the maintenance of an orderly relation of exchange.[5]

The liberal practicality shied away from any utopian conception of the public domain and was content with an order that seemed to emerge through nonintervention in the natural processes, or rather in the metabolism, of society. As Hannah Arendt has argued, this extraordinary identification of society with its economy may be traced in part to the liberal devaluation of politics:

> What concerns us in this context is the extraordinary difficulty with which we, because of this development, understand the decisive division between the public and private realms, between the sphere of the polis and the sphere of household and family, and, finally, between activities related to a common world and those related to the maintenance of life, a division upon which all ancient political thought rested as self-evident and axiomatic.[6]

All the spaces of the modern world are absorbed into a single economy whose rhythms are linear and mechanical. The architecture of public and commercial institutions, the furnishings of the home, and even the styles in which we clothe our bodies, threaten to destroy the dialectic between the things that are to be shown and the things that are to be hidden. Le Corbusier has called the modern house "a machine to live in," a machine that mechanizes living in a mechanical world. In a strange, disordered repetition of ancient symbolism, the modern household is hooked into the center of the universe through its television navel and suspended by an aerial (*universalis columna quasi sustinens omnia*) between heaven and hell. "The Kwakiutl believe that a copper pole passes through the three cosmic levels (underworld, earth, sky); the point at which it enters the sky is the 'door to the world above.'" The visible image of this cosmic pillar in the sky is the Milky Way.[7] Through the picture-frame windows of the modern house the metabolism of family life is projected into the public realm and from there it completes its circuit back into the home through a magical aether populated by waxes, deodorants, soap-suds and tissues:

> To behold, use, or perceive any extension of ourselves in technological form is necessarily to embrace it. To listen to radio or to read the printed page is to accept these extensions of ourselves into our personal system and to undergo the "closure" or displacement of perception that follows automatically. It is this continuous embrace of our own technology in daily use that puts us in the Narcissus role of subliminal awareness and numbness in relation to these images of ourselves. By continuously embracing technologies, we relate ourselves to them as servo-mechanisms. That is why we must, to use them at all, serve these

objects, these extensions of ourselves, as gods or minor religions. An Indian is the servo-mechanism of his canoe, as the cowboy of his horse, or the executive of his clock.[8]

In one-dimensional society desire is no longer domesticated. Now the whole of society is organized to satisfy domestic passions. And this is an arrangement eminently suited to the ethic of individualistic familism and the socialization of the members of society into their "calling" as consumers whose needs are the self-imposed agency of social control. It is this continuity of psychic and socioeconomic space which grounds the coherent fantasy of consumer sovereignty at the same time that it fills the air with the noise and filth that are the by-products of the commercial narcosis to which we now turn our attention in so-called political ecology, or the politics of the environment.

METABOLISM AND POLITICAL ECONOMY

In the period between the decline of the feudal family order and the rise of modern nation-states geared to a fully industrialized economy, there emerged a microcosmic version, in the Court circle and the salon of high society, of the tragic alienation of the individual in a universe hidden from God and abandoned to the play of social forces.[9] Whether it is through the identification of the individual with his title at Court in the *ancien régime* or with his occupational status in the modern corporation, the modern individual encounters a bureaucratization of private sensibilities, a wasteland between the boundaries of the heart and the public presentation of the self.[10] The rise of modern society involves the history of the decline of feudal community, the growth of the nation state, industrial technology, and political democracy. But it also generates the paradox of the affinity of individualism for conformism through the erosion of the communal bases of the family, the guild, the village, and the Church.[11] The emergence of the "total community" has its origins in the growth of rationalism in economics, politics, and religion. In each of these areas, modern individualism receives its impulse from the subjectivization of the bases of the feudal community and a simultaneous assimilation of the individual into the abstract community of market society:

> Thus, from the viewpoint of this enlightened political economy which has discovered the *subjective* essence of wealth within the framework of private property, the partisans of the monetary system and the mer-

cantilist system, who consider private property as a *purely objective*
being for man, are *fetishists* and *Catholics*. Engels is right, therefore, in
calling Adam Smith the *Luther of political economy*. Just as Luther recognized *religion* and *faith* as the essence of the real *world* and for that
reason took up a position against Catholic paganism; just as he annulled
external religiosity while making religiosity the *inner* essence of man;
just as he negated the distinction between priest and layman because he
transferred the priest into the heart of the layman; so wealth external to
man and independent of him (and thus only to be acquired and con-
served from outside) is annulled. That is to say, its *external* and *mindless objectivity* is annulled by the fact that private property is incorpo-
rated in man himself, and man himself is recognized as its essence. But
as a result, man himself is brought into the sphere of private property,
just as, with Luther, he is brought into the sphere of religion. Under the
guise of recognizing man, political economy, whose principle is labour,
carries to its logical conclusion the denial of man. Man himself is no
longer in a condition of external tension with the external substance of
private property; he has himself become the tension-ridden being of private property. What was previously a phenomenon of *being external to
oneself*, a real external manifestation of man, has now become the act
of objectification, of alienation. This political economy seems at first,
therefore, to recognize man with his independence, his personal activity, etc. It incorporates private property in the very essence of man, and
it is no longer, therefore, conditioned by the local or national *characteristics of private property* regarded as existing outside itself. It manifests
a cosmopolitan, universal activity which is destructive of every limit
and every bond, and substitutes itself as the *only* policy, the *only* universality, the *only* limit and the *only* bond.[12]

The identification of the metabolism of the household—whose
place is stimulated by the media's claim to inform and service its needs—
with the national economy, which results in the hybrid concern of "political
economy," is the outcome of the alienation of private property and labor
from their anchorages in use-values. In their endlessly reproducible forms,
as the exchange-values of capital and labor power, private property and
labor enter the public realm and subordinate the public realm to the needs
of market society. The emancipation of labor is the precondition of the substitution of exchange-values for use-values which leads to the subordination
of all fixed forms of life and property to the accumulation and expansion of
wealth. In the remarkable passage from the *Economic and Philosophical
Manuscripts* quoted above, Marx explains how private property becomes
the subjective impulse of industrial activity through its definition as *labor-
power*. The Physiocrats identified all wealth with land and cultivation, leav-

ing feudal property intact but shifting the essential definition of land to its economic function and thereby exposing feudal property to the later attacks on ground rent. The objective nature of wealth was also in part shifted to its subjective basis in labor, inasmuch as agriculture was regarded as the source of the productivity of land. Finally, industrial labor emerged as the most general principle of productivity, the factors of production, land, labor, and capital being nothing else than moments in the dialectic of labor's self-alienation.

PRIVATE OPULENCE AND PUBLIC SQUALOR

Marx's expectation that capitalism would collapse because of the conflict between the technological rationalization of its economy and the irrationality of its social and political structure remains unfulfilled. The question is whether the Marxian diagnosis is as irrelevant as the phenomena of welfare and affluent capitalism[13] are taken to suggest. Certainly, the metabolism of the corporate economy absorbs more than ever the public and private energies of modern society. Under the banner of a neofeudal ideology of corporate responsibility,[14] a new psychic serfdom to brand-loyalties and occupational status immunizes monopoly capitalism from the processes of social and political criticism. It is increasingly difficult to discuss the nature of the good society where everyone is mesmerized by the *goods* society. Here political vision recedes as tele-vision expands to the limits of the globe which it simultaneously delimits.

In the North American context, there are historical and environmental factors which contribute to the equation of politics and abundance.

> The politics of our democracy was a politics of abundance rather than a politics of individualism, a politics of increasing our wealth quickly rather than dividing it precisely, a politics which smiled both on those who valued abundance as a means to safeguard freedom and on those who valued freedom as an aid in securing abundance.[15]

The ideological roots of the affluent society have been traced by John Kenneth Galbraith to the hold upon the liberal mind of certain imperatives which flow from the "conventional wisdom." Adam Smith, Ricardo, and Malthus were clear enough that the mass of men were powerless against the class of property owners. But in view of the factors of scarcity, against which any proposal for social redistribution could only mean a relapse into barbarism, it seemed that the mutual interests of the rich and the poor lay in the expansion of industrial activity. However, in the conventional wisdom the imperative of production remains just as imperious as it ever was, despite intervening changes in the modern economic environment which have made abundance a technological possibility, if not a sociological certainty:

> These—productivity, inequality and insecurity—were the ancient pre-occupations of economies. They were never more its preoccupations than in the nineteen thirties as the subject stood in a great valley facing, all unknowingly, a mountainous rise in well-being. We have now had that mountainous rise. In a very large measure the older preoccupations remain.[16]

The paradox of the affluent society is that it has exhausted the liberal imagination in a "solution" of the problems of inequality and insecurity through a mindless expansion of production and consumption. The instrument of this paradoxical situation is the corporate organization[17] of the economy whose success in controlling its economic environment has won for it political acceptance from its employees and the stabilizing support of state-administered antidepressants for those moments in which the soulfulness of the corporation threatens to reach a low point. The power of the corporation to control its environment assumes a variety of forms, ranging from its ability to control price-cost relationships, levels and composition of investment, the nature of research and innovation, the location of industry with its effects upon local communities and, of course, its power to influence governmental intervention, and, last but not least, its mediatized power to shape the physical and sociopsychological environment of the consumer public. In each case, these powers of the corporation are of enormous social and political consequence.[18]

In the face of the reality of corporate power, Galbraith's theory of *countervailing power* is hardly more than a figleaf for corporate respectability and liberal prudishness. It is in any case a desperate gesture in view of Galbraith's own understanding of the corporate practice of integrating its production and sales efforts through the generations of wants. By engineering consumer response, the corporation is able to get an *ex post facto* ratification of its commitment of social resources as determined by the corporate agenda. While paying lip service to consumer sovereignty in the final allocation of social resources, the corporation can in fact assume the conventional distribution of social resources between the public and private sectors of the economy. This presumption is a political reality inasmuch as the demand for public services presently arises out of the needs of low-income groups who are powerless to compete away social resources from the private uses of higher-income groups. It is only in the context of the unequal distribution of income, which remains as much as ever a defining characteristic of affluent capitalism, that one can properly understand the imperative of production or, rather, of *relative overproduction for the private sector*, which in turn promotes the secondary imperatives of consumption and other-directedness. Despite the heralds of the age of high mass consumption, the fact is that monopoly capitalism is a production system continually faced with the problem of deficient consumption structurally related to the class distribution of income.[19] Because of this conventional restraint upon the *economic space* of the capitalist system, it is necessary to invade the *psychic space* of workers and consumers through raising levels of expectation or through deepening levels of credit.

There are, of course, attempts to expand the economic space of the capitalist system through extensions of the public sector, overseas operations, and the conquest of outer space. But in no case do these extensions result in a significant alteration of the flow of social resources between the public and private sectors. The commanding position of the corporation in the face of governmental efforts to redistribute social income is evident from the relative stability of corporate profits after taxes as a share of national income.[20] In effect, the government merely uses the corporation to collect its taxes and is therefore dependent upon the corporate economy's agenda having been substantially realized before it can undertake its own program. Indeed, it must be recognized that the determination of the balance between the public and the private sectors of capitalist society depends heavily on the identification of welfare and warfare, as well as the conquest of outer space. The American war psychosis is an obvious manifestation of the militarized production imperatives of the corporate economy. The significance of the social imbalance created by military spending is lost when considered simply as a proportion of total gross national product. From this perspective, the amount of GNP absorbed in military expenditure seems negligible and easily enough absorbed in alternative expenditures. However, once it is realized that military expenditures represent half of total federal government expenditures,[21] it is clear that the issue is neither negligible nor easily corrigible. It is not negligible because it represents the impoverished conception of the public domain in capitalist society. Nor is it easily corrigible, since to find alternative paths of governmental spending involves a reconsideration of the balance between the public and private sectors which would expose the poverty of the liberal ideology.

The institution of advertising as a *specular function* can now be understood as the essential means of expanding the economic space of capitalism in a manner compatible with the liberal ideology. What Galbraith calls the "dependence effect" is in reality a political option which, if unrelated to the class structure of capitalist society and its effects upon the distribution of social resources between public and private uses, appears as the myth of an evil genius:

> Were it so that a man on arising each morning was assailed by demons which instilled in him a passion sometimes for silk shirts, sometimes for kitchenware, sometimes for chamber pots and sometimes for orange squash, there would be every reason to applaud the effort to find the goods, however odd, that quenched the flame. But should it be that his passion was the result of his first having cultivated demons, and should it also be that his effort to allay it stirred the demons to even greater and greater effort, there would be question as to how rational was his solution. Unless restrained by conventional attitudes, he might wonder if the solution lay with more goods or fewer demons.
>
> So it is that if production creates the wants it seeks to satisfy, or if the wants emerge *pari passu* with the production, then the urgency of the wants can no longer be used to defend the urgency of the production. Production only fills a void that it has itself created.[22]

It is not the dependence effect as such which is responsible for the irrationality of consumer behavior. For in every society wants are largely cultural acquisitions. The real problem is the nature of the social order which determines the content and pattern of wants. A society which fails to maintain the necessary complementarities between private and public goods and services drives itself even deeper into the accumulation of private amenities in order to compensate for the public squalor which this very process leaves in its wake.

The automobile becomes the true symbol of the North American flight into privacy.[23] It has hollowed the cities and drained the countryside, melting each into the atomized living-space of suburbia; it is the instrument of urban congestion and rural uglification. At the same time, the automobile is perfectly geared to the values of technical rationality, private ownership, individual mobility, sex equality, and social rivalry—preeminently the values of the liberal ideology and the stock-in-trade of the corporate economy. The automobile is eminently the equilibrator of the tensions in corporate culture: It is a family headache and a family joy, an air-pollutant indispensable for trips into the fresh air of the countryside, an escape mechanism from all the problems with which it is structurally integrated. The role of the automobile in modern society makes it evident that we can no longer consider machines from the purely technological standpoint of the mastery of nature. We must take into account the interaction between machinery and the social relations between men not only in the context of machine production but in the wider context in which machinery patterns our style and ecology of life. Short of such an understanding, we find ourselves hallucinating the conquest of distance while all the time the road which opens up before us is the distance between a humane living-space and the little boxes which house our automobiles.

But any criticism of the automobile is likely to be dismissed as quixoticism. For the power to respond to such criticism has been sapped through the cultivation of psychic identification with the automobile as an extension of individual personality. Even where slightly less elongated extensions have been preferred by four-wheeled man, the apparent rationality of that choice actually only deepens the commitment to private as opposed to public transportation alternatives. The result is a chain-process in the privatization of other social resources integrated with the automobile culture at a time when more than ever we need to break that circuit.

The difficulty of intervention on behalf of the public domain is nowhere better seen than in the light of the potential hue and cry against interference with the individual's freedom to buy, own, and drive, wherever and whenever, that capsule which seals us off from physical and social reality while making us completely dependent upon it. Likewise, the course of public intervention in regard to the automobile is indicative of the impoverished conception of government in liberal society. The result so far is the confusion of the growth of public space with the extension of public highways which breeds more automobiles and accelerates the dislocation of urban spaces in favor of suburban locations. Commuting by means of private transportation becomes the only link between living spaces and working spaces. Finally, the rationale of this living arrangement is given a coherent projection through television advertisements in which the enjoyment of suburban values can be "seen" in the happy use of the automobile to take children to school, mother to the stores, father to work, and the dog to the veterinarian, without anyone even wondering how everything got so far away. The hard work in all this is now evident with the appearance of the family van—or recreation vehicle (RV) and the new van mothers.

ALIENATION AND THE SUBLIMATION OF POLITICS

There is a trend in industrial society towards the interpretation of freedom and equality in terms of consumer behavior rather than as political action about the nature and conditions of production and consumption. "Equality for the working-classes, like freedom for the middle-classes, is a worrisome, partially rejected, by-product of the demand for more specific measures."[24]

In the context of corporate capitalism, the rhetoric of freedom and equality no longer swells into a coherent political ideology as it once did as a strategy of bourgeois and proletarian emancipation. Just as the terrible freedom of market society has not always been tolerable to the middle class without escapes, so the working-class response to market society has varied from class struggle to the becalmed acceptance of inequality softened by improvements to the social basement. This ambivalence in the response to the symbolism of freedom, equality, and reason must be understood in terms of the changing social contexts from which these nations derive their meaning and significance. Robert A. Nisbet has commented upon the changing contexts of individuality.[25] He observes that when we speak of "the individual" we are dealing with an ideal type or moral abstraction whose symbolic currency depends upon the existence of an institutional context which is favorable to its assimilation in everyday life. The liberal image of man, its possessive individualism, is the result of the imputation of the properties of market society to the interior life of the individual. The liberal theory of society and the individual was plausible just so long as the historical situation which liberalism presupposed effectively linked its vocabulary of motives with typical contexts of action.[26] However, once the evolution of market society moves in the direction of corporate society, the vocabulary of liberalism merely evokes lost contexts, arousing a nostalgia haunted by the loss of meaning.

The loss of a meaningful, social, or public context for the ideals of individualism, freedom, and equality is reflected in the alienated and confused symbolism of David Riesman's *Lonely Crowd* or Paul Goodman's *Growing Up Absurd*. Each of these works confronts us with the paradox that society may be free without individuals being free. The liberal identity of individual and social interests, or, rather, the liberal perception of the challenge and opportunity offered to the individual by society, has withered away into a conviction of the absurdity of society and the idiocy of privatization which is its consequence. For want of a genuine public domain, in which the political and social activities of individuals can achieve a focus and historical perspective, men abandon politics for the civic affairs of suburbia or the "bread and butter" questions of unionism. By shifting awareness toward improvements in consumption styles, these tactics deflect attention from the social imbalance which results from the pursuit of intraclass benefits that leave whole sectors of the population outside of their calculus. This tactic is further strengthened by the ideological acceptance of social improvement through the escalation effect of an expanding economy upon all classes rather than through any radical redistribution of class income or the extension of chances of individual mobility between classes.

As individual awareness is increasingly shifted toward a concern with consumption, economic knowledge is reduced to a concern with prices in abstraction from the corporate agenda which determines prices. The result is a loss of any coherent ideological awareness of the political and economic contexts of individual action. However, this does not represent an end of ideology. It is simply the nature of the dominant ideology of individualism shaped by the context of corporate capitalism. The irony here—as is evident from "The Price is Right"—is that the poor generally have a very shaky idea of prices. The power of the American ideology is visible in the way that the contestants treat their knowledge/ignorance of prices as moral knowledge—for which they are rewarded, luckily or unluckily, by the great "give-away," as on *Who Wants To Be a Millionaire*.

In order to break the tendency to monetize all individual experience, and in order to shift individual time perspectives away from short-term consumer expectations, it is necessary to institutionalize more universal goals of collective and long-term value. Such a requirement falls outside the pattern of instant satisfactions projected by the consumer orientation. The latter substitutes the thin continuity of progress for the solid accumulation of social history. The result is the paradox upon which Robert E. Lane has commented. "Is it curious," he asks, "that a nation that has so emphasized progress should have no sense of the future? I do not think so," he replies, for "progress is a rather thin and emotionally unsatisfactory continuity. It is the continuity of differences, the regularity of a rate of change, almost a rate of estrangement."[27] Any concern with social balance, institutional poverty, and waste, or the interaction between politics, economics, and culture presupposes a collective and historical framework; but this is foreign to the liberal ideology of individual agency and its moralistic acceptance of inequality and failure within a natural order of social competition and private success, portrayed everyday in its quiz and game shows.

In a society where individual interests are so privatized that people fear and denigrate public activities, common effort is likely to be viewed only as a substitute for private effort. Moreover, any comparison between public and private enterprise will be moralized in favor of the fruits of individual effort owing to the very real struggle involved in the acquisition of private pools, homes, and education. The loss of community functions resulting from the privatization of social resources makes individual accumulation appear all the more "rational." In reality, the individual is driven toward this pattern of privatization not from genuine choice but because he is deprived of alternatives whose systematic provision would require a public sector powerful enough to compete with the private economy. The provision of alternatives to the patterns of production and consumption in the corporate economy never gets beyond the platitudes of "variety" and consumer sovereignty which are virtually meaningless once attention is diverted from increasing the size of the goods basket to questions about the quality of a single item in it. The rise of parasocial, political, and economic activities is an indication of individual withdrawal from "society." It is the expression of an abstracted individualism that is the ideological alternative to political action on behalf of a world that men can have in common. This loss of a common world separates society into a corporate hierarchy and a multitude of individuals who are turned in upon themselves in the competition to maintain occupational status and at the same time other-directed in their attempt to rationalize their loss of community in the pursuit of the good life—family-style. Where there is no common world between working life and private life the individual's public life is reduced to shopping expeditions, church attendance, and movie-going, all homogenized to suit family tastes, which are, of course, presensitized to the appeal of the "goods life."

It is in keeping with the liberal ideology of *individualistic familism* that tensions are personalized and at best call for individual therapy. Any attempt to relate private troubles to institutional contexts, which would suggest public or political action, is regarded as projection, the evasion of difficulties best tackled within the four walls of the home, whether in the TV room or the bedroom. The result is that men and women lack bridges between their private lives and the indifference of the publics that surround them:

> Nowadays men often feel that their private lives are a series of traps. They sense that within their everyday worlds, they cannot overcome their troubles, and in this feeling they are often quite correct. What ordinary men are directly aware of and what they try to do are bounded by the private orbits in which they live; their visions and their powers are limited to the closeup scenes of job, family, neighbourhood; in other milieux, they move vicariously and remain spectators. And the more aware they become, however vaguely, of ambitions and of threats which transcend their immediate locales, the more trapped they seem to feel.[28]

It is the task of the political and sociological imagination to conceive men's private troubles in the contexts of public concern and to furnish bridging concepts which will enable individuals to translate their private uneasiness into public speech and political action. It must undertake to shift the contexts of freedom, equality, and reason away from the private sector and out of the household into a public domain which will constitute a genuine common world. And this is a task which must be articulated in a conception of government which is bold enough to seek understanding and responsible control over the human and social values generated but largely dissipated in the corporate economy which enforces the privatization of men's lives. Such a positive conception of government would help to create a public domain in which men and women share common assumptions about their moral and physical environment and exercise them in a concern for truth of speech and beauty of form in public places—places cleared of their present monuments to financial cunning and the fear of the future that wastes private lives in reality—and on TV—ad nauseam.

NOTES

1. Herbert Marcuse, *One-Dimensional Man: Studies in the Ideology of Advanced Industrial Society* (Boston: Beacon Press, 1964). For the distinction between technical or "functional" rationality and "substantial" rationality, see Karl Mannheim, *Man and Society in an Age of Reconstruction* (London: Routledge and Kegan Paul, 1940), pp. 51-60.

2. "*To be socially integrated in America is to accept propaganda, advertising and speedy obsolescence in consumption.* The fact that those who fit the image of pluralist man in pluralist society also fit the image of mass man in mass society. Any accurate picture of the shape of modern society must accommodate these ambiguities." Harold L. Wilensky, "Mass Society and Mass Culture: Interdependence or Dependence?" *American Sociological Review* 29:2 (April 1964):196.

3. Marcuse, *One-Dimensional Man*, pp. 10-11. For an empirical confirmation of the ideological content of the consumer orientation, see Sanford M. Dornbush and Lauren C. Hickman, "Other-Directedness in Consumer-Goods Advertising: A Test of Riesman's Historical Typology," *Social Forces* 38:2 (1959):99-102.

4. Aristotle, *Politics*, 1252 a.2.

5. C.B. Macpherson, *The Political Theory of Possessive Individualism: Hobbes to Locke* (Oxford: The Clarendon Press, 1962), p. 3. The liberal concept of "society" provoked the counterconcept of "organic society" in Conservative and Marxian thought which have more in common than either has with liberalism. Karl Mannheim, *Essays on Sociology and Social Psychology* (New York: Oxford University Press, 1953), Chap. II, "Conservative Thought."

6. Arendt, *The Human Condition* (Chicago: University of Chicago Press, 1958), p. 28.

7. Mircea Eliade, *The Sacred and the Profane* (New York: Harcourt, Brace and Company, 1959), p. 35.

8. Marshall McLuhan, *Understanding Media. The Extensions of Man* (New York: McGraw Hill, 1965), p. 46.

9. Lucien Goldmann, *Le dieu caché: Etude sur la vision tragique dans les penseés et dans le théatre de Racine* (Paris: Librarie Gallimard, 1955).

10. Locke shows no awareness of the alienation of man in society, unlike Hobbes, who, nevertheless, has no solution for it. It is Rousseau who first attempts to link the experience of alienation with social criticism.

11. Robert A. Nisbet, *Community and Power* (New York: Oxford University Press, 1962). Cf. Karl Marx, *Communist Manifesto* (Gateway Editions, 1954), p. 12. Marx's sketch of the breakdown of feudalism is brilliantly developed in Karl Polanyi, *The Great Transformation* (Boston: Beacon Press, 1957).

12. *Karl Marx: Early Writings*, trans. and ed. by T.B. Bottomore (London: C.A. Watts and Company, 1963), pp. 147-148.

13. For a careful appraisal of the relations between welfare capitalism and the affluent society, see T.H. Marshall, *Sociology at the Crossroads* (London, 1963), Part Three, "Social Welfare," and Richard M. Titmuss, *Essays on the "Welfare State"* with a new Chapter on "The Irresponsible Society" (London: Allen and Unwin, 1963).

14. It has been argued that the corporate exercise of political power is in principle continuous with the natural-law tradition of the separation of sacred and profane power and its institutionalization in the countervailing powers of feudal nobility. St. Augustine's "City of God," understood as the theory that in every age there is a moral and philosophical framework which constrains power, has been claimed as the model of corporate politics. Adolf A. Berle, *The Twentieth Century Capitalist Revolution* (London: Macmillan, 1955).

15. David M. Potter, *People of Plenty: Economic Abundance and the American Character* (Chicago: University of Chicago Press, 1954), p. 126.

16. J.K. Galbraith, *The Affluent Society* (Boston: Houghton Mifflin Company, 1958), p. 77.

17. "By and large, corporations have been able to exert sufficient pressure on governments, and on social institutions generally, to stabilize the field in their favour. *This stabilizing of the environment is the politics of industry.*" J. Porter, *The Vertical Mosaic: An Analysis of Social Class and Power in Canada* (Toronto: University of Toronto Press, 1965), p. 269 (my italics). The weakness of government planning in America has been attributed to the competitive nature of its political institutions which weaken it relative to the more monolithic structure of business. A. Shonfield, *Modern Capitalism: The Changing Balance of Public and Private Power* (New York: Oxford University Press, 1965), p. 353.

18. Carl Kaysen, "The Corporation: How Much Power? What Scope?" and Norton Long, "The Corporation, Its Satellites and the Local Community," in *The Corporation in Modern Society*, Edward S. Mason, ed. (Cambridge, MA: Harvard University Press, 1961), pp. 85-105, 202-217.

19. Gabriel Kolko, "The American 'Income Revolution'," *America as a Mass Society*, in Philip Olson, ed. (New York: Free Press of Glencoe, 1963), pp. 103-116.

20. Irving B. Kravis, "Relative Shares in Fact and Theory," *American Economic Review* (December 1959):1931, quoted in Paul A. Baran and Paul M. Sweezy, *Monopoly Capital: An Essay on the American Economic and Social Order*

(New York: Monthly Review Press, 1966), p. 148. There is consistent empirical evidence of long-run tax shifting by corporations. E.M. Lerner and E.S. Hendrikson emphasize the long-run maintenance of a stable after-tax rate of return on investment despite substantial increases in corporation income tax in "Federal Taxes on Corporate Income and the Rate of Return in Manufacturing 1927-1952," *National Tax Journal* IX (September 1965):193-202; R.E. Slitor, "The Enigma of Corporate Tax Incidence," *Public Finance* XVIII (1963):328-329. Very little changes by updating these figures.

21.   In a record U.S. budget of $135 billion for the fiscal year 1967, $72.3 billion were allocated to military expenditures, a sum exceeded only by the figure of $81.3 billion spent in 1945. *Globe and Mail*, 25 January, 1967, p. 1; in 1986 the U.S. military budget was 300 billion, or 7 per cent of GNP. See Stephen Gill and David Law, *The Global Political Economy: Perspectives, Problems, and Policies* (Hemel Hempstead: Harvester, 1988). These figures are easily updated but generally ignored.

22.   J.K. Galbraith, *The Affluent Society*, p. 153.

23.   Housing would illustrate the problems of overprivatization and the impoverishment of the public sector just as well as the automobile to which it must be related. The housing situation is especially illustrative of the tendency to privatize even explicitly public functions.

24.   Robert E. Lane, *Political Ideology: Why the American Common Man Believes What He Does* (New York: Free Press of Glencoe, 1962), p. 60.

25.   *Community and Power*, chap. 10, "The Contexts of Individuality," Compare C.B. Macpherson's discussion of "social assumptions," *The Political Theory of Possessive Individualism*, chaps. 1, 2, 3, 6.

26.   C. Wright Mills, "Situated Actions and Vocabularies of Motive," in *Power, Politics and People: The Collected Essays of C. Wright Mills*, I.L. Horowitz, ed. (New York: Ballantine Books, 1963), pp. 439-452.

27.   *Political Ideology*, p. 290

28.   C. Wright Mills, *The Sociological Imagination* (New York: Grove Press, 1961), p. 3.

CHAPTER THREE

The Political Economy of Narcissism: Some Issues in the Loss of Family *Eros*

In North America the contemporary discussion of eroticism, narcissism, and intimacy reflects the rejection of the radical version of Eros and the body-politic that flourished in the 1960s. The vehemence of the denunciation of New Left Politics is all the more remarkable when one considers its abandonment of the hopes that the Left once inspired. We shall argue that the embittered critics of the culture of narcissism (Lasch)[1] and of the politics of intimacy (Sennett)[2] fail to differentiate the historical vicissitudes of *Eros* and its political economy. They promote an oversimplified contrast between the realms of public and private space, ignoring earlier work by Arendt,[3] Marcuse,[4] and O'Neill.[5] A peculiar feature of the reduction of *Eros* to the negative practices of eroticism is its lament for the loss of family and its expropriation by liberal professional and welfare agencies. With remarkable prescience, these trends were pointed to by Rieff as the next stage of American culture:

> Where family and nation once stood, or Church and Party, there will be hospital and theatre too, the normative institutions of the next culture.

55

Trained to be incapable of sustaining sectarian satisfactions, psychological man cannot be susceptible to sectarian control. Religious man was
born to be saved; psychological man is born to be pleased. The difference was established long ago when "I believe", the cry of the ascetic,
lost precedence to "one feels", the caveat of the therapeutic. And if the
therapeutic is to win out, then surely the psychotherapist will be his
secular spiritual guide.[6]

The full sense of this observation involves us in an extended analysis of arguments regarding the symbiosis between the politics of intimacy,
the sexualization of economic life, and the mystification of the bases of
social control and power in advanced capitalist society. These are, of course,
"seen" but unanalyzed in the media and, as we are generally arguing, this
specular function of the media serves to deepen our ignorance and to extend
our social amnesia. By the same token, it makes analysis "difficult" for the
average citizen.

In order to widen the present critical context, we shall examine
Habermas' early thesis[7] on the relation between the bourgeois family in the
generation of critical public opinion. We do so to reformulate the issues
raised by Lasch and Sennett in terms of a critical theory of the legitimation
problem in the context of political democracy. We shall then return to the
critical importance of Marcuse's concept of *memory* in defense of the New
Left legacy which must be preserved despite the current will to amnesia and
the *postmodern repression of all utopian politics*.[8]

NARCISSISM AND DEFAMILIZATION

It is a commonplace of contemporary social analysis that our troubles stem
from ourselves. The great achievements of bourgeois individualism are not
nearly so faulted by their failed universalism—or in comparison with the
present realities of Socialist individualism—as by the souring of individualism into egoism and its parasitic demands upon genuine sociability, independence, and critical judgment. The voices in this complaint are legion.
They are as varied as the talk of parents, teachers, priests and journalists
paraded by the media. We propose to look at two laments upon the collapse
of *Eros* into what Lasch, for one, calls *narcissism* and what Sennett, for the
other, calls *intimacy*. Since the laments of Lasch and Sennett are in part
motivated by disenchantment with the infantile disorders of the 1960s New
Left (*Delenda est Chicago*), we take a closer look at the Freudian and
Marxist legacy that spawned these misfits so curiously left over by the com-

mercialization of their seemingly utopian demands. In order to avoid the limitations of an insufficiently historicized conception of the political foundations of the relations between the state and society, and between the economy, family, and individual, we review the history of these relationships. The benefit of these two moves is that we may draw proper distinctions between *Eros* as a *critical* and *regenerative* force, as opposed to its historical collapse into sexuality, or narcissism and intimacy. At the same time, we may draw *Eros* out of a purely utopian space into the real political history of the family and the liberal state.

We begin with Lasch's *The Culture of Narcissism*, although it appeared after Sennett's *The Fall of Public Man*, to which it addresses specific criticisms. Rather than summarize either work, we present each respective to the other only from the standpoint of the problematic of their sensitivity to the historical vicissitudes of the cultural malaise they refer to respectively as "narcissism" and "intimacy." Since all of the authors with whom we are dealing are writing within the matrix of Freudo-Marxist analysis, we might stress from the outset that, inasmuch as their common focus is the process of social reproduction, it is the *family* that constitutes their principal concern; and it is only from this standpoint that we pursue the vicissitudes of *Eros*. For this reason, we consider Lasch's argument to be found not so much in his "bestseller" observations upon narcissism—their generality in fact providing for the very readership they condemn—but in his concern with the displacement of the socialization functions of the bourgeois family onto professional, bureaucratic, and state agencies.[9] It is these agencies that foster the narcissistic culture that is generated in the reduced families whose main function is now consumption (aided by TV viewing) rather than production. The narcissistic personality is the perfect expression of the weakened relation of the family vis-à-vis the state and economy which recruit only privatized consumers. Thus Lasch argues that the schools, juvenile courts, health and welfare services, advertising and the media all function to erode the authority of the family. The result is that the family is increasingly a place where narcissistic individuals learn to compete with one another in the consumption of services and goods—emotional, political, and economic—but with a diminished capacity for the competence required in their production:

> The psychological patterns associated with pathological narcissism, which in less exaggerated form manifest themselves in so many patterns of American culture—in the fascination with fame and celebrity, the feat of competition, the inability to suspend disbelief, the shallowness and transitory quality of personal relations, the horror of death— originate in the peculiar structure of the American family, which in turn

originates in changing modes of production. Industrial production takes the father out of the home and diminishes the role he plays in the conscious life of the child. The mother attempts to make up to the child for the loss of the father, but she often lacks practical experience of child-rearing, feels at a loss to understand what the child needs, and relies so heavily upon outside experts that her attentions fail to provide the child with a sense of security. Both parents seek to make the family into a refuge from outside pressures, yet the very standards by which they seek to bring it about, derive in large part from industrial sociology, personnel management, child psychology—in short, from the organized apparatus of social control.[10]

Lasch does not stop to specify the extent to which the scenarios he projects largely represent the media specularization of an articulate and overexposed middle-professional family milieu in which the absence of father and the dominance of mother are serviced by high incomes that enable such families to trade in narcissism to an exaggerated degree. But what should not be overlooked in Lasch's generalizations is the core argument that the collapse of the family and the bureaucratic fostering of narcissistic personality is the instrument of the mystification of the sources and operation of modern domination:

Therapeutic forms of social control, by softening or eliminating the adversary relation between subordinates and superiors, make it more and more difficult for citizens to defend themselves against the state or for workers to resist the demands of the corporation. As the ideas of guilt and innocence lose their moral and even legal meaning, those in power no longer enforce their rules by means of the authoritative edicts of judges, magistrates, teachers, and preachers. Society no longer expects authorities to articulate a clearly reasoned, elaborately justified code of law and morality; nor does it expect the young to internalize the moral standards of the community. It demands only conformity to the conventions of everyday intercourse, sanctioned by psychiatric definitions of normal behaviour.[11]

In short, the seriousness of Lasch's argument lies not in its analytic reach but in the hypothesis which we shall reformulate as the *refeudalization of the private realm under welfare liberalism*. The environmentalist vocabulary of the social sciences—that is, the reduction of responsibility to causation and environmental victimization—is a prime agency in weakening the responsibility of the individual and the authority of the family on behalf of the administrative control of the state and corporate economy. In addition, as we have argued elsewhere, *the medicalization of social control*[12] compounds these effects. Moreover, this is a form of social control into which individuals are willingly coopted through the expansion of carnal ignorance, the channels it provides for their egoism and narcissistic impotence:

> Having overthrown feudalism and slavery and then outgrown its own personal and familial form, capitalism has evolved a new political ideology, welfare liberalism, which absolves individuals of moral responsibility and treats them as victims of social circumstance. It has evolved new modes of social control which deal with the deviant as a patient and substitute medical rehabilitation for punishment. It has given rise to a new culture, the narcissistic culture of our time, which has translated the predatory individualism of the American Adam into a therapeutic jargon that celebrates not so much individualism as solipsism, justifying self absorption as "authenticity" and "awareness."[13]

This state of affairs is aided equally by the forces of progressivism, liberalism and conservatism. Worse still, Left radicalism has no better insight into the issues and its criticism of the family combined with its own political narcissism has only aggravated the very problem it claims to remedy.

Although Lasch is a historian, the central weakness of his *Culture of Narcissism* is not its successful pop sociology but its anachronizing argument that is its peculiar historical weakness. Yet Lasch is at pains in his opening chapter to bring the same charge of anachronism against Sennett's *The Fall of Public Man*, even though the latter superficially outsteps Lasch at his own trade.[14] What is the issue, then, that I am referring to as an *anachronizing argument*? Quite simply, it is the problem of trying to locate within history a stable point for the evaluation of that same history into progressive and regressive periods. It is a problem that is especially acute in the reconstruction of history to service emancipatory critical theory. We shall be more concerned with it once we turn to Habermas' analysis of the public realm, apart from our earlier remarks in Chapter 2. Lasch expands his thesis on the culture of narcissism largely as a rejection of the political culture of the last decade or so of American society adjusting to the passing of the 1960s radicalism. He separates himself from similar social critics, and in particular from Sennett, on the grounds that their attacks upon narcissism only weaken further the grounds of genuine individualism and family privacy. Sennett's thesis that the collapse of the boundary between the private and public worlds undermines a rational politics based upon enlightened self and class interest, is dismissed as an ideological exaltation of bourgeois liberalism and civility—as though the latter were immune to their own mystification of their irrational domination of society. Yet, when Lasch himself looks for a counter-culture to narcissism, the best he himself can do is to locate it in the hard school of the very rich, realistic about privilege and victimization, busy in the pursuit of studies, music lessons, ballet, tennis and parties "through which the propertied rich acquire discipline, courage, persistence, and self-possession."[15]

Lasch's nostalgia for the old rich and his contempt for the new managerial elites and professionals who replace them, is charitably to be understood as a lament for the loss of family. It expresses his belief in the family as the only ground of a responsible individualism capable of forming into "communities of competence"[16] that in turn might form a bulwark against the modern administrative state. Such communities might draw their energies from a creative, material *Eros*, as envisaged by Marcuse and Norman O. Brown.[17] But Lasch cuts himself off from this vision because he has turned sour on the New Left (as well as feminism and ecologism) without attempting to understand what was originally intended by its efforts to link politics and *Eros*.[18] We shall return to this task later, once we have taken a closer look at Sennett's argument.

THE POLITICS OF INTIMACY

In *The Fall of Public Man*—which is curiously silent about the work of Arendt—whom we have discussed in Chapter 2 and of Habermas to whom we shall turn—Sennett claims that contemporary American society is characterized by a lack of interest in impersonal issues that are constitutive of basic social and political processes and by vociferous interest only when these issues are, quite mistakenly, treated as issues of sentiment and personality. It is argued that this is the result of the erosion of the boundaries between public and private space which formerly kept intimacy in its proper place and left society and politics to their own practices of public order, decorum, and the pursuit of rational self-interest. The staging area of this historical division of sentiment and rationality was the bourgeois family:

> The modes of public and private expression were not so much contraries as alternatives. In public the problem of social order was met by the creation of signs; in private the problem of nurturance was faced, if not solved, by the adherence to transcendental principles. The impulses governing the public were those of will and artifice; the impulses governing the private were those of restraint and the effacement of artifice. The public was a human creation, the private was the human condition.[19]

We shall pass over Sennett's historical narrative since the ground it surveys is better covered in Habermas' historical study of the public realm, the vicissitudes of public opinion and of modern publicity, on which we shall comment later. Instead, we shall draw from Sennett's argument only those threads which, like those in Lasch's theory of narcissism, seem to go towards the more specific claim that narcissism and intimacy as standards for the judgment of the public process function essentially to mystify the political order, while gratifying the illusion of pseudocommunity (destructive *gemeinschaft*). Such a community does not function to foster political action. Because it derives from the collectivization of personality, it can only engage in communal purification:

> People are seeking for others to disclose themselves in order to know where to belong, and the acts of disclosure consist in these details which symbolize who believes what, rather than what should be believed. Baring of a self becomes the hidden agenda of political life. And when, in fantasy, these details revealing who is fighting are then blown up to stand for a collective person, political community becomes moralistic rather than ideological in tone.[20]

It should be noticed, however, that Sennett and Lasch come to opposite conclusions about the consequences of narcissism and intimacy. Whereas Lasch sees these forces as fostered by the State, Sennett believes that they overwhelm the State. Increasingly, the politics of intimacy restrict people to life in the family, the school, and the neighborhood at the expense of the city and the state. Such a contradictory conclusion reflects the superficiality of both arguments, though they can be reconciled if we see that, as Habermas, Foucault,[21] and O'Neill[22] have argued, civil privatism is fostered by the State in order to give sway to its own administrative intervention in the economy, health, education, and welfare sectors that shape the contours of family and individual life. The price of such privatism, in Sennett's view, is that men become incompetent with civility and urbanity. They lose the power to pursue their rational self-interest, to preserve a proper distance between themselves and the agony of competition, victimization, and exploitation. Politics are ruled by personality, charisma, and the media mobilization of opinion which heightens the general absorption in motive and personality politics.[23] The less men have to do with one another—because in fact they are ruled by impersonal administrative processes—the more they are seen to be involved with one another through *the illusion of political intimacy*. Finally, Sennett, like Lasch, is equally hostile to his own earlier encounters with the political inexperience of radical community organization. Local participation is no substitute for the hard experience of challenging the basic structure of power in the state and economy. In Sennett's view, the only place to apprentice ourselves to such knowledge and action is in the school of the city where men aggressively pursue their interests.

THE HOMOGENIZED SOCIETY AND THE FAMILY OF EROS

So far we have considered two overlapping theses on the collapse of *Eros* into narcissism and intimacy, each of which leads to an alienated stand on the possibility of a politics and community generated by *Eros* as a positive and critical force. We have subordinated criticism of Lasch and Sennett— especially of their distortions of the intellectual tradition they trade upon— to letting their accounts stand as a framework for the North American discussion of sexuality. In particular, we have stressed that narcissistic sexuality is discussed as a mystifying ideology of political intimacy that is both the creation of and response to the impersonal administration of society and political economy. Narcissism is diagnosed as the social and cultural malaise produced by the decay of the bourgeois family which had provided a reality principle for liberal individualism. Given the obvious but ignored similarity between the more serious treatment of their principal theses in the work of Arendt, Marcuse and Habermas,[24] we should now introduce the reservations that Lasch and Sennett carelessly set aside.

In *The Human Condition* (1958) Hannah Arendt opened the discussion of the consequences for political thought and action that derive from the processes of social homogenization of labor and property achieved through modern political economy. She points out that this is an historical shift achieved through the disembedding of the "life processes" of the private household economy. Here slavery and painful necessity were the rule but were classically regarded as a prepolitical condition of public life and the pursuit of freedom. What is peculiar to modern times is the invention of a realm between the public domain and the private society which threatens to ruin both of these spheres. Thought and speech are thereby deprived of an enduring public realm and so become reduced either to social chatter or to the manipulative vocabularies of the administrative and behavioral sciences, as Marcuse[25] and O'Neill[26] have shown. Political economy is therefore the hybrid science of a mass slave society beyond anything envisaged in the ancient world and without the restraining realm of the genuinely political sphere where men and women pursue excellence:

> The social realm, where the life-process has established its own public domain, has let loose an unnatural growth, so to speak, of the natural; and it is against this growth, not merely against society but against a constantly growing social realm, that the private and the intimate, on the one hand, and the political (in the narrower sense of the word), on the other, have proved incapable of defending themselves.[27]

Here, then, Arendt sees clearly what Lasch and Sennett leave unanalyzed. The central fact of modern society is the *homogenization* of both the public and private domains which has left intimacy without a vital shelter and subordinated political conduct to administrative expertise.[28] Neither Lasch nor Sennett, in turning their backs upon the New Left, wish to consider how narcissism and intimacy could possibly have been part of the political struggle against the homogenization of modern society and personality.

At this point, we really cannot postpone consideration of the principal theses in Marcuse's *Eros and Civilization* (1955, 1962 paperback). We recognize, of course, that Marcuse has often enough been regarded as the prophet of the narcissistic culture abhorred by Lasch, Sennett, and their general public.[29] Nor do we mean to overlook that Marcuse, Habermas, and Adorno indeed suffered at times from an infantile interpretation of their teachings. But the risks in their case were no greater than the neoconservative risks in the laments of Lasch and Sennett.[30] The latter two completely eviscerate Marcuse's analytic contribution to modern social psychology by emphasizing only the negative side of reading Freudian instinct theory into political economy. The crux of Marcuse's contribution is his *rapprochement* between Marx's concept of *economic surplus* and Freud's theory of instinctual repression and his formulation of a two-stage theory of historically necessary and historically needless *surplus repression*. Any form of society requires a modification of instinctual expression in view of the basic necessity of work and the division of labor. Under advanced capitalism, however, this historical stage of the reality principle, enshrined in the ethic of performance and competitive achievement, might be set aside in favor of the release of a positively creative and communal *Eros*. Instead, advanced capitalism reaches even higher levels of surplus repression masked as the narcissistic pursuit of highly sexualized objects and scenarios of commodity consumption.

Marcuse makes it clear that it is the homogenized administrative processes of the economy and society that collapse the family space within which a mature ego might develop with a capacity for authentic nonconformity rather than an aimless, autistic, and aggressive secondary narcissism. Whereas previously the oedipal relations of the family at least supported a conflict between generations, the modern family presents no face to its children:

Now, however, under the rule of economic, political and cultural monopolies, the formation of the mature superego seems to skip the stage of individualization: the generic atom becomes directly a social atom. The repressive organization of the instincts seems to be *collective*, and the ego seems to be prematurely socialized by a whole system of extrafamilial agents and agencies. As early as the preschool level, gangs, radio, and television set the pattern for conformity and rebellion; deviations from the pattern are punished not so much within the family as outside and against the family.[31]

Lasch and Sennett ignore that in the 1960s—as once more in this millennium—it was the children of these families who turned away from the administrators of the school, home, university, state, and corporations—who demanded their palpable presence in scenes of real confrontation and to music of peace, kindness, and love. To have dismissed all this in the 1970s and 1980s, merely because of its aberrations, is to foreshorten political memory and to contribute irresponsibly to the collective amnesia that reinforces the politics of communicative and intellectual incompetence.[32]

It is part of the historical record that the work of Marcuse and of Norman O. Brown—despite their differences—aroused the political awareness of North America and Western Europe. By bringing together Marx and Freud, the bodily costs of modern economic and political life became apparent, despite the seeming affluence and happiness of its own self-advertisements. Marcuse made clear the interaction between the levels of instinctual and institutional repression while rereading Freudian metapsychology in order to historicize the struggle between Eros and Thanatos and to reveal the politically constructive norms of love's body. Brown, of course, went further, dissolving Marcuse's civilization of property and family in a Blakean vision of the resurrected body. To interpret these excesses, O'Neill revived the classical concept of the *body politic* as the primordial basis of all political life, as the flesh that must suffer and resist bioindustrial and sociopolitical oppression.[33] Brown's vision of revolutionary politics as political delinquency may well have transcended the working demands of the world's family. The same is probably true of Hoffman and Rubin's street theatre. Yet it cannot be denied that the children of the 1950s taught the family of the 1970s the language of love and gentleness, if only the family knew how to hold out against the professionals waiting to finish it off.

So far from encouraging narcissism in any vulgar sense, Marcuse in fact drew upon the mythological and aesthetic tradition[34] of Narcissus and Orpheus pitted against the tradition of Prometheus:

> The Orphic and Narcissistic experience of the world negates that which sustains the world of the performance principle. The opposition between man and nature, subject and object, is overcome. Being is experienced as gratification, which unites man and nature so that the fulfillment of man is at the same time the fulfillment, without violence, of nature. . . . This liberation is the work of Eros. The song of Orpheus breaks the petrification, moves the forests and the rocks—but moves them to partake in joy.[35]

It is in terms of this aesthetic myth that Marcuse then adapted the concept of *primary narcissism* which he interpreted not as a neurotic symptom but as a constitutive element in the construction of reality and a mature, creative ego with the potential for transforming the world in accordance with a new science of nature:

> The striking paradox that narcissism, usually understood as egotistic withdrawal from reality, here is connected with oneness with the universe, reveals the new depth of the conception: beyond all immature autoeroticism, narcissism denotes a fundamental relatedness to reality which may generate a comprehensive existential order.[36]

Marcuse, like all the critical theorists—and despite Lasch and Sennett—never imagined that the culture of primary narcissism could emerge from the degraded sexual culture that has arisen out of the collapse of the bourgeois family. Like all genuine countercultural notions, Marcuse's *civilization of Eros* presupposes a dialectical confrontation of society with the real authority of the family. As Horkheimer observed, it is only in the family that the external processes of impersonalization can be resisted. Despite society's inroads upon its member's relations, only in the family is there a real chance to foster human beings for their own sake:

> If it is true that family life has at all times reflected the baseness of public life, the tyranny, the lies, the stupidity of the existing reality, it is also true that it has produced the forces to resist these. They flashed forth when the child hung on his mother's smile, showed off in front of his father, or rebelled against him, when he felt someone shared his experiences—in brief, they were fostered by that cozy and snug warmth which was indispensable for the development of the human being.[37]

For all its limitations, Horkheimer nevertheless regarded the family as the source not only of authority but of a countervailing vision of "a better condition for all mankind." The risk in this strikingly sentimental observation is that it could be taken by Lasch, for example, as nothing but a piece of the conventional sociological wisdom that insists upon educing the family to a "haven in a heartless world."[38] This would often mean that the critical theory of the family, after having first pointed to the twin pressure upon family life of narcissism and the administrative state, finally succumbed to the therapeutic ideology and its undialectical separation of public and private life. We need therefore to reopen the analysis of the role of the bourgeois family in the division of public and private space, as I have done elsewhere.[39] For whatever else that is said about the contemporary ideological functions of the temporary sexualization of consumption, our understanding of the politics of intimacy and the therapeutic ideology of the liberal welfare state turns upon an adequate account of the underlying crisis in motivation and legitimation in liberal democratic states.

In his *The Structural Transformation of the Public Sphere*, Habermas is concerned with the historical and formal preconditions of individual rights in a democratic society. He therefore analyses the historical role of the bourgeois family in the mediation of individual rights between the state and civil society. Where the family is eroded, public opinion is swamped by the media politics of personality, economic life is masked in the eroticization of commodity consumption, and democracy is threatened with the refeudalization of the public agenda. In an effort to avoid both the Hegelian synthesis of state and society, which sacrificed participatory democracy, and the Marxist reduction of formal law to mere ideology, Habermas attempts to reconstruct the liberal concept of public opinion as the normative institutional context for the pursuit of critical truth:

> The self-awareness of the political realm demonstrated in the fundamental category of a legal norm is transmitted through the institutional awareness of the literary public. These two formations of the public are strangely intertwined. In both cases it happens that a public realm of private persons, whose autonomy rests on the disposal of private property, seeks to represent itself as such within the sphere of the bourgeois family, in love, freedom and culture; in a word, as humanity.[40]

The historical function of bourgeois public space as the realm of critical public opinion lay in its opposition both to the absolute state (monarchy) and the sheer interest of (civil) society. This critical sphere originated in the private sphere of the literary salons and cafes which fostered free speech, and the civil irrelevance of persons and status in the pursuit of critical discussion. Moreover, political speech was not, as Arendt required, totally divorced from any substantive interest in the realm of necessity. Bourgeois society was, for a time at least, able to subject its social and economic interests as well as those of the state to the broader concerns of a rational and humanitarian public:

> If ideologies do not simply show the falseness of socially shaped consciousness; if, moreover, they possess a momentum which raises existence beyond itself in a utopian way—if only for legitimation purposes—then, to the extent this is true, ideology first appears from this time. Its origin would lie in the identification of the "proprietor" with "man," both in the role assigned to private persons as audience in the public sphere of the bourgeois state (in the identification of the political and literary public) and in the role of public opinion itself where class interest, expressed through public debate, can assume the appearance of universality—in the identification of domination with its transformation into pure reason.
>
> Whatever the case, the developed bourgeois public is tied to a complex constellation of social conditions. These have always undergone deep and rapid changes. These changes reveal the paradox of the institutionalized public realm in the bourgeois state. Its basic principle, opposed in its very nature to any form of domination, helped to establish a political order whose social basis could not after all render domination superfluous.[41]

Thus, contrary to Lasch and Sennett, Habermas locates humanity and intimacy in the private sphere of the bourgeois family as potentially emancipatory norms of critically rational public life—and not as escapist ethics. Rather than being swamped by the private realm—that is, the modern realm of political economy—it is the state which has politicized bourgeois public opinion through the formation of political parties, a politicized press and media. Thus the formal legitimacy of modern democratic society—despite obvious contradictions with its underlying socioeconomic inequalities—is able to live off the surplus-value of an ideal of a critical rational communicative community of individuals who in practice are privatized. Lasch and Sennett see this but cannot explain it. Rather, they seize upon commonplaces of the consequences of this development, scandalously subverting their intellectual sources and making nothing of the issue of the normative foundations of political democracy in the welfare state. Here the superiority of Habermas' historical analysis of the active functions of public opinion and the processes that threaten it is evident:

> The political public sphere of the social welfare state is characterized by a peculiar weakening of its critical functions. At one time the process of making proceedings public (*Publizität*) was intended to subject persons or affairs to public reason, and to make political decisions subject to appeal before the court of public opinion. But often enough today the process of making public simply serves the arcane policies of special interests; in the aim of "publicity" it wins public prestige for people or affairs, thus making them worthy of acclamation in a climate of nonpublic opinion. The very words "public relations work" (*Oeffentlichkeitsarbeit*) betray the fact that public sphere must first be constructed case by case, a public sphere which earlier grew out of the social structure. Even the central relationship of the public, the parties and the parliament is affected by this change in function.[42]

Lasch and Sennett grasp only the most superficial aspects of the role of the degradation of *Eros* in the politics of intimacy and the eroticization of the consumer life. They sense that sexuality contributes to the legitimation of consumerism while simultaneously increasing family dependence upon the therapeutic ideology of liberal welfare capitalism. But, having turned away from the New Left and any positive appraisal of Frankfurt critical theory, they refuse to raise any analytic questions regarding the *normative grounds of political democracy*. Despite the tendency of critical analysis to swamp itself in the one dimensionality thesis, Habermas has nevertheless struggled to elaborate both an abstract linguistic ideal of the communicative bases of participatory democracy and an historical reconstruction of the normative institutional framework of critical public opinion. It is vital to critical theory that these two dimensions—the analytic and historical—be kept together in specific studies of the processes of political democracy.[43] The issue is whether bourgeois democracy has spent its moral capital to the point where the forces for homogenized administrative culture can no longer be resisted except by fundamentalist and neoconservative elements. If this is so, then the culture of narcissism, commodity sex, and the politics of intimacy represent the narcosis of family authority and authentic individualism. Despite appearances, the fun culture of sovereign political and economic consumers only releases the passions in order more effectively to fetishize and homogenize them:

> The most important motivation contributed by the socio-cultural system in advanced capitalist societies consists of syndromes of civil and

familial vocational privatism. Civil privatism here denotes an interest in the steering and maintenance (*Versorgung*) performances of the administrative system but little participation in the legitimizing process. . . . Civil privatism thus corresponds to the structures of a depoliticized realm. Familial vocational privatism complements civil privatism. It consists in a family orientation with developed interests in consumption and leisure on the one hand, and in a career orientation suitable to status competition on the other. This privatism thus corresponds to the structures of educational and occupational systems that are regulated by competition through achievement.[44]

It is therefore interesting to see Habermas arguing that where the administrative process seeks to take over the formerly self-legitimating culture of the schools, health, the city, and the family, it reopens critical discussion of these domains, destroying the administrative split between the public and private realms. But this means that the moral capital of bourgeois capital is shifted from being a dead legacy of the state into the realm once more of critical public discourse:

The stirring up of cultural affairs that are taken for granted thus further the politicization of areas of life previously assigned to the private sphere. But this development signifies danger for the civil privatism that is secured informally through the structures of the public realm. Efforts at participation and the plethora of alternative models—especially in cultural spheres such as school and university, press, church, theatre, publishing, etc.—are indicators of this danger, as is the increasing numbers of citizens' initiatives.[45]

It is significant that, unlike Marcuse, Habermas does not include in the cultural spheres just listed either science or industry. Nor has he much to say on the media. He throws out of hand any ideas of a "new science" because, despite his critique of positivism, Habermas seems not to consider the natural science basis of industry and technology to be open on its own level to the norms of critical discourse.[46] This apart, we must surely recognize the moral capital of the New Left in any further discussion and innovation with respect to the issues of participatory democracy. It is too easy to banish the New Left with unhappy memories of student excesses without any analysis of the symbiosis between administrative rationality and popular irrationality in scenes of confrontation.[47] Moreover, it was the New Left that tried to take the aesthetic revolt into the streets in order to realize the subversive intent of art which Habermas recognizes as part of the emancipatory legacy of the bourgeoisie. In the end, although retreatism is always a response that youth may adopt, revolt is its other response. The political task, however, is to find ways to bring together student and working class modes of political reaction with those of citizens whose interests are less easily identified with those of youth, women, and workers, although all sense they need some common way of dealing with the administration of their lives. Today, the proletariat is hardly the icon of their common plight, any more than the rhetorical tropes of socialism can be presumed to provide the necessary complex analytic grasp of modern political processes.

For these reasons, I have tried elsewhere[48] (see Chapter 5) to extend the classical image of the *body politic*. To capture the cultural politics of the 1960s, I distinguish the *libidinal* body politic from the *productive* body politic and the *bio-body* politic. The latter two levels are those upon which we pursue the needs of the working and family body. The rhetoric of the levels of work and family, as well as the analytic sciences appropriate to them, differs from the aesthetic rhetoric of the libidinal body politic—and will appear more prosaic in comparison to that of love's body. The task is to find ways of integrating the values and institutional requirements of this trilevel conception of the body politic. In this task, we shall need to create communities of discursive competence capable of confronting the professional administration of the formerly self-regulating spheres of the bodypolitic. To the extent that we are capable of this, it is likely that we shall find ourselves trading upon the religious and moral capital we are otherwise encouraged to treat as the dead weight of the past. Critical theorists have no other answer to the sources of the emancipatory ethic they impute to western history. Nor is there any.[49]

In summary, then, the contemporary discussion of narcissistic sexuality is largely a lament for the loss of family functions. This is so because the liberation of the individual from the family has not resulted in the emancipation of *Eros*. To account for this, it has been necessary to rethink the historical functions of the bourgeois family. Such studies have revealed that the stripping of the bourgeois family only released its members into the professional, therapeutic, and administrative care of the liberal welfare state.[50] This is the hinge that replaces the bourgeois family. At the same time, the administrative and therapeutic state is ideologically suited to mask the processes of social control that respond to and direct the lives of individuals raised in *families without authority*. The loss of the bulwark function of the bourgeois family, and the consequent hybridization of the public and private realms into "the social," represents the loss of an institutional basis for the role of critical public opinion in the legitimating process of liberal political democracy. Liberalism may well be the spent ideology of late capitalism, having lost its basis both in the family and the nation state.[51]

Finally, the destruction of family authority is also at the root of the loss of *memory* (*anamnesis*) which, as Marcuse has argued so eloquently, is the ability to preserve a historical and critical perspective upon the issues of justice and injustice as the fundamental human bond. This, above all, is what is lost if we lose the legacy of Marcuse's thought. As I see it, memory is not a practical technique, however much it is valued in the meager tasks of school, work, and public dimension of *Eros*; it is the insistence that good prevail over evil in everyone's life not only in our own. *Memory* is, therefore, family-and-future regarding; it is the will to recapture for future society every past happiness. By the same token, memory is not to be confused with nostalgia, or narcissistic dreaming. Rather, historical memory is our collective will to a civilized utopia. It is, therefore, resolutely opposed to current postmodern amnesia:

> This ability to forget—itself the result of a long and terrible education by experience—is an indispensable requirement of mental and physical hygiene without which civilized life would be unbearable; but it is also the mental faculty which sustains submissiveness and renunciation. To forget is also to forgive what should not be forgiven if justice and freedom are to prevail. Such forgiveness reproduces the conditions which reproduce injustice and enslavement: to forget past suffering is to forgive the forces that caused it—without defeating these forces. The wounds that heal in time are also the wounds that contain the poison. Against this surrender to time, the restoration of remembrance to its rights, as a vehicle of liberation, is one of the noblest tasks of thought. In this function, remembrance (*Erinnerung*) appears at the conclusion of Hegel's *Phenomenology of the Spirit*; in this function, it appears in Freud's theory. Like the ability to forget, the ability to remember is a product of civilization—perhaps its oldest and most fundamental psychological achievement. Nietzsche saw in the training of memory the beginning of civilized morality—especially the memory of obligations, contracts, dues. This context reveals the one-sidedness of memory-training in civilization; the faculty was chiefly directed toward remembering duties rather than pleasures; memory was linked with bad conscience, guilt, and sin. Unhappiness and the threat of punishment, not happiness and the promise of freedom, linger in memory.[62]

Much of the 1960s has gone to waste. But this is true of every cultural decade. Yet each may add something to the other provided we do not think of the future as so very much different from what we have always been. The same is true of the time between youth and old age. Of course, this charitable regard is not strong in contemporary culture. We are as determined to separate the past from the future as we are to project the future ahead of our present. The separation of the decades—no less than the separation of the generations—serves to undermine our cultural memory. As such it serves to deepen the political amnesia favored by administration politics and its preference for a docile citizenry whose sense of time and space is collapsed in the self-centerdness of mass culture. Today we nevertheless draw upon the legacy of the 1960s for our reorientation towards militarism, bureaucracy, racism, and sexism. The struggle is no longer uniquely identified with youth. Other groups and movements have extended the struggle—and these in turn will be replaced because the struggle will be a long one. We should therefore remember that each age owes to its near past. Such memory can only strengthen the future.

NOTES

1. Christopher Lasch, *The Culture of Narcissism: American Life in an Age of Diminishing Expectations* (New York: Warner Books, 1979); *The Minimal Self: Psychic Survival in Troubled Times* (New York: W.W. Norton and Company, 1984).

2. Richard Sennett, *The Fall of Public Man* (New York: Alfred A. Knopf, 1977).

3. Hannah Arendt, *The Human Condition* (Chicago: The University of Chicago Press, 1958).

4. Herbert Marcuse, *Eros and Civilization: A Philosophical Inquiry Into Freud* (Boston: Beacon Press, 1955).

5. See Chapter Two.

6. Philip Rieff, *The Triumph of the Therapeutic: Uses of Faith After Freud* (London: Chatto and Windus, 1966), pp. 24-25.

7. Jürgen Habermas, *Strukturwandel der Offentlichkeit, Untersuchungen zu einer Kategorie der bürgerlichen Gesellschaft* (Neuweid and Berlin: Hermann Luchterhand Verlag, 1962); now *The Structural Transformation of the Public Sphere: An Inquiry into a Category of Bourgeois Society*, translated by Thomas Burger (Cambridge: MIT Press, 1989).

8. For a review, see Richard Guarasci, *The Theory and Practice of American Marxism 1957-1970* (Lanham: University Press of America, 1980); Peter Clecak, *Radical Paradoxes: Dilemmas of the American Left 1945-1970* (New York: Harper and Row, 1973); Richard King, *The Party of Eros: Radical Social Thought and the Realm of Freedom* (Chapel Hill: University of North Carolina Press, 1972).

9. *The Culture of Narcissism* has generated considerable comment. This is easily available in "A Symposium: Christopher Lasch and the Culture of Narcissism," *Salmagundi* 46 (Fall 1979); and "Special Symposium on Narcissism," *Telos* 44 (Summer 1980). Our own argument tries to avoid repetitions to be found in the preceding commentary. See also note 29 below. For a feminist commentary on Lasch, see Jessica Benjamin, "Authority and the

Family Revisited: Or, a World Without Fathers?" *New German Critique,* 13 (1978):35-57; and her "The Oedipal Riddle: Authority, Autonomy, and the New Narcissism," in *The Problem of Authority in America,* J.P. Diggins and M. Kann, eds. (Philadelphia: Temple University Press, 1981), pp. 195-224.

10. *The Culture of Narcissism,* pp. 301-302.

11. Ibid., pp. 315-316.

12. John O'Neill, "Sociological Nemesis: Parsons and Foucault on the Therapeutic Disciplines," in *Sociological Theory in Transition,* Mark L. Wardell and Stephen P. Turner, eds. (Boston: Allen and Unwin, 1986), pp. 21-35.

13. *The Culture of Narcissism,* pp. 369-370.

14. But only superficially. Sennett's use of history and his self-serving notions of dialectical and hermeneutical method have been severely criticized. See J.H. Plumb, "When Did Citizens Become Strangers?" *The New York Times Book Review,* January 23, 1977, p. 3; and Sheldon Wolin, "The Rise of Private Man," *The New York Review of Books* (April 14, 1977), pp. 19-26.

15. *The Culture of Narcissism,* p. 371.

16. Ibid., p. 396.

17. Norman O. Brown, *Love's Body* (New York: Vintage Books, 1966).

18. John O'Neill, "Gay Technology and the Body Politic," in *The Body as a Medium of Expression,* Jonathan Benthall and Ted Polhemus, eds. (New York: E.P. Dutton, 1975), pp. 291-302.

19. *The Fall of Public Man,* p. 98.

20. Ibid., p. 238.

21. Michel Foucault, *The History of Sexuality,* Volume I (New York: Vintage Books, 1980).

22. John O'Neill, "The Disciplinary Society: From Weber to Foucault," *British Journal of Sociology* XXXVII/1 (March 1986):42-60.

23. See Chapter Ten.

24. Ben Agger, "On Happiness and the Damaged Life," in *On Critical Theory,* John O'Neill, ed. (New York: The Seabury Press, 1976), pp. 12-33.

25. Herbert Marcuse, *One Dimensional Man: Studies in the Ideology of Advanced Industrial Society* (Boston: Beacon Press, 1964), Ch. 4, "The Closing of the Universe of Discourse."

26. O'Neill, "Sociological Nemesis."

27. *The Human Condition,* pp. 43-44.

28. For Habermas' critical view of Arendt's concept of the political realm, see Jürgen Habermas, "Hannah Arendt's Communications Concept of Power," *Social Research* 44/1 (Spring 1977):3-24. For a spirited reply, see David Luban, "On Habermas on Arendt on Power," *Philosophy and Social Criticism* 6/1 (Spring 1979):79-95.

29. Lasch's amnesia with respect to Marcuse and Brown is remarkable. They are not even given a reference in *The Culture of Narcissism.* But see *The Minimal Self,* pp. 227-240.

30. I am aware that Lasch considers it a misreading to treat his work as a lament. See his "Politics and Social Theory: A Reply to Critics," *Salmagundi* 46 (Fall 1979):194-202.

31.   *Eros and Civilization*, p. 88.

32.   Agger, "On Happiness and the Damaged Life."

33.   John O'Neill, "Authority, Knowledge and the Body Politic," in *Sociology as a Skin Trade* (London: Heinemann, 1972), pp. 68-80.

34.   Shirley Sugerman, *Sin and Madness: Studies in Narcissism* (Philadelphia: The Westminster Press, 1976).

35.   *Eros and Civilization*, p. 151.

36.   Ibid., p. 27. It is striking that Norman O. Brown later reformulated Freud's theory of narcissism in a similar fashion. See his "The Self and the Other: Narcissus," in *The New Eroticism: Theories, Vogues and Canons*, Philip Nobile, ed. (New York: Random House, 1970), pp. 9-23.

37.   Max Horkheimer, *Critical Theory: Selected Essays*. Trans. by Matthew J. O'Connell et al. (New York: Herder and Herder, 1972), p. 277. See also Phil Slater, *Origin and Significance of the Frankfurt School: A Marxist Perspective* (London: Routledge and Kegan Paul, 1977), pp. 107-112.

38.   Christopher Lasch, *Haven in a Heartless World: The Family Besieged* (New York: Basic Books, 1977), p. 92.

39.   John O'Neill, *Five Bodies: The Human Shape of Modern Society* (Ithaca: Cornell University Press, 1985); *The Missing Child in Liberal Theory: Towards a Covenant Theory of Family, Community, Welfare, and the Civic State* (Toronto: University of Toronto Press, 1994).

40.   *Strukturwandel der Offentlichkeit*, p. 73, my translation.

41.   Ibid., p. 111, my translation.

42.   Jürgen Habermas, "The Public Sphere: An Encyclopedia Article (1964)," *New German Critique* 1/2 (Fall 1974):55.

43.   See Chapter Five.

44.   Jürgen Habermas, *Legitimation Crisis*. Trans. by Thomas McCarthy (Boston: Beacon Press, 1975), p. 75.

45.   Ibid., p. 72.

46.   John O'Neill, "Marxism and the Two Sciences," in his *For Marx Against Althusser, And Other Essays* (Washington, D.C.: University Press of America, 1982), pp. 151-175.

47.   For a corrective view see John O'Neill, "On Body Politics," in *Recent Sociology* No. 4, "Family, Marriage and the Struggle of the Sexes," Hans Peter Drietzel, ed. (New York: The Macmillan Company, 1972), pp. 251-267.

48.   O'Neill, *Five Bodies*.

49.   John O'Neill, *The Communicative Body: Studies in Communicative Philosophy, Politics and Sociology* (Evanston, IL: Northwestern University Press, 1989).

50.   Jacques Donzelot, *The Policing of Families*. Trans. by Robert Hurley (New York: Pantheon Books, 1979); see Chapter 4.

51.   Herbert G. Reid, "American Liberalism, Authority, and the Corporate State: A Critical Interpretation." *Annual Review of Political Science*, vol. 3 (1990):134-159.

52.   *Eros and Civilization*, p. 212.

CHAPTER FOUR

Women as a Medium of Exchange: Defamilization and the Feminization of Law in Early and Late Capitalism

Since the family is at once the setting and the subject of so much of the communication industry's messages that specularize the changes in its structure and values with respect to gender relations, intergenerationality, consumption, health and community, it is necessary to situate "the family" in a broad historical and political perspective.[1] If we are to have any critical stance upon the general process of defamilization that now seems irreversible, we must try to release ourselves from the daily specularization of its effects which now furnish one of the basic staples of the media.

We necessarily engage in an argument with the feminist perspective upon these events. But since feminism is varied in its knowledge and practices, we shall start from a central text in Levi-Strauss whose interpretation provides a convenient source for the parting of the ways. The text is valuable since it formulates a grand law of communication, namely, that all things must be regarded as exchangeable against one another and that the exchange of women is the prototype of economic and linguistic exchange. But we should have it in Levi-Strauss's words:

[General laws of exchange] can be achieved only by treating marriage regulations and kinship systems as a kind of language, a set of processes permitting the establishment, between individuals and groups, of a certain type of communication. That the mediating factor, in this case, should be the *women of the group*, who are *circulated* between clans, lineages, or families, in place of the *words of the group*, which are *circulated* between individuals, does not at all change the fact that the essential aspect of the phenomenon is identical in both cases.[2]

Levi-Strauss himself comments upon the claim that this law of communication is an antifeminist law. He points out that, of course, women are not reducible, like words, to signs without value. Precisely because women are producers of speech their fate is not sealed by any structuralist principle of communication, yet nor can feminist discourse, as Juliet Mitchell has very well argued in this connection, emancipate itself by reducing all social constraints to a manipulable sociology or technology.[3]

I shall argue that whereas in early capitalism the family was moved into the factory and legislated out in favor of *defamilism* and *feminism*, today, feminist and defamilist discourse serve to legislate the factory into the family in the interests of late capitalism. In either case, there is an alliance between legal-medical discourse, feminism and defamilism, as particular discursive strategies which must be analyzed historically if we are to understand modern *biopolitical economy*. I think it can be shown that legal-medical discourse functions to feminize and defamilize individual bio-rights. The latter extend the fiction of the juridico-legal subject, while eroding correlative familized duties except as these are mediated through the agencies of the liberal welfare and therapeutic state.[4]

I am not going to develop this argument in all of the detail that it requires. Rather, I shall present an argument-sketch whose only justification is that it may serve to provide some conceptual orientation in a realm where there is considerable confusion. I do not believe that this is the case simply because we are confused analysts. Rather, I think that proponents of the liberal bourgeois state, of defamilism and individualism, have developed multiple and frequently bipolar discursive strategies for the transformation of the role of the individual, the family, and the state in the shift from early to late capitalism. In trying to analyze these shifts, I am offering a study in the spirit of Foucault's work, as well as drawing upon work of my own regarding the *political economy of the body*.[5]

Since I start from a structuralist or familist perspective, I shall begin with some remarks upon the family economy and then try to specify the changing legal status of the *family wage* and the *social wage* in the contexts of early and late capitalism. In the course of doing; so, I shall be critical of what I call *feminist economics*, as well as of *feminist psychoanalysis*, since I think them to be conceptually inadequate and detrimental to the defense of the family. I cannot see how there can be any profound grasp of the human rights and duties of men, women, and children that is not grounded in a proper understanding of our familied being.

SOME ISSUES IN FEMINIST ECONOMICS

Every society must reproduce itself on a number of levels. The issue is whether the same logic of reproduction prevails at each level. Only a rather crude understanding of Marx would lead us to think so—and yet Marx may not be blameless here. Thus we might make at least the following distinctions when thinking of social reproduction as it bears upon the family:

1. Bio-reproduction
2. Everyday living
3.. Socialization
4. Work
5. Law and order

Surely all of these tasks are productive in the sense that they are necessary functions of any society. To consider work productive and the rest unproductive because they do not exchange on a market is peculiar. Worse yet, since some of these functions do in some respects exchange on a market, they are then both productive and unproductive. But is a woman who sells her milk more productive, more individualized and emancipated than a mother breast feeding her own child? Or is a husband who supports his wife in the latter case exercising his ownership of her body in order to indulge his preference for breast-fed children rather than send his wife's milk to market, or her with it? If there is any ideology in keeping the mother at home, then the same is surely to be said of the case where she goes to market. The freedom to earn wages for looking after children, feeding and caring for them, is exchanged against a certain rate of exploitation. Because there are mothers at work canning peas who need day-care services there must be mothers at work in day-care service who have to buy canned peas. Of course, the actual exchange is much more roundabout—but at every point in the cycle surplus value is extracted from the work process.

To purchase the labour-power of a family of four workers may, perhaps, cost more than it formerly did to purchase the labour-power of the
head of the family, but, in return, four days' labour takes the place of
one, and their price falls in proportion to the excess of the surplus
labour of four over the surplus labour of one. In order that the family
may live, four people must now, not only labour, but expend surplus-
labour for the capitalist. Thus we see that machinery, while augmenting
the human material that forms that principal object of capital's exploit-
ing power, at the same time raises the degree of exploitation.[6]

To argue that a number of family functions, carried out by wives
and mothers, are more rational when exchanged on the market is to argue
for the strangest of things—namely, the rationality of the market. In
Weberian terms, it is to confuse the *formal* (exchange) *rationality* of the
market with the quite different question of its *substantive* (ethical, human)
rationality. One has only to contrast the family as a substantively rational
context of love and care with the market and the professions as a context for
these "goods," or "services." Even the crudest Marxist, or feminist, should
see the socioeconomic and sociopsychological difference between a wife
and a prostitute, or between a mother and a day-care worker. Things are no
different when we view the claim that the logic of the family and the market
is peculiarly suited to males. Through work all men are subject to the logic
of production in the narrow economic sense. It is structurally naive to argue
that the logic of production and the market place is governed by a *male
logic*—unless we are to believe that men have discovered in work the best
outlet for their inherent sado-masochism. At work men are disciplined,
timed, degraded, exploited, injured, aged and often killed. They are so
because they are at work in order to be exploited. When it is difficult to
exploit them, they are unemployed. When they refuse to be exploited they
are either on strike, or shut-out. Underlying all of this, and increasingly lost
in current discussion, is the fact that men and women are familied beings
who divide between them (however they do so) the tasks of reproducing
themselves as average members of the society in which they live, subject to
its legal and political economy.

If men earn a *family wage*, as I shall argue in more detail later, it is
spent as a family wage. The husband who is possessive about his wife's
hair-curlers, or his daughter's nail varnish, or the baby's diapers, is as
unusual as the husband who drinks up his pay packet. Only family chaos
could result from a husband requiring industrial and commercial criteria of
productivity and pace in the household economy of kitchen and bedroom. In
fact, only feminists could imagine such defamilizing tactics as part of their
war upon the male economy they so misconceive. *Feminists are ideologists.*

That is to say, they believe that the family economy is a slave economy and should be replaced by a free economy of working and consuming individuals whose only ties are legal, contractual relations. They consider that the logic of contract is sufficient (or justifiably expensive) in each of the domains of special reproduction outlined above to achieve this shift. In this, they display a more coherent vision of capitalist society than it has so far practiced. They often do so in the name of socialism and humanism, and while tolerating in themselves a temporary middle-class advantage. Feminists have declared the family an economic and psychological ghetto in exchange for careers they hope will buy off their own bodies and marriages, or else make them necessary to the families they reject and degrade.

The defamilization of the household is advanced by what we may call *feminist economics*. The latter turns upon an incongruity procedure which involves costing and status ranking women's work in the home as though it were wage labor in the market. Even home sex has been viewed in this light, no less than mothering and a variety of housework tasks. On all fronts domestic labor does not score highly! Yet it is argued that the monetization of these services is desirable if their true value is to be recognized. In part, this is implicit in government and insurance practices that compute costs to these services, or whenever a given household is obliged to purchase some or all of them in the market. Curiously enough, this reasoning leads to a double and contradictory conclusion. Thus *most men cannot afford their wives, while society cannot afford not to charge itself for their services*—housewives, like petroleum, presumably needing to command world prices if we are to have a rational economy. Among other arguments, it can be suggested that levels of education are important forces bearing upon a woman's marketability. It is ignorant to stay out of the market. Moreover, it is immoral. Women ought not to be uniquely leisure consumers. It is better for them to work for their leisure, to learn in order to work, and to let men into the nonwork, nonmarket uses of living. Thus the Equal Rights Amendment sought to degender work qualifications, although presumably it could not get around the female work of hosting the embryo and birthing—even on the minimalist version of reproduction.

The fact is, of course, that such feminist economics are hopelessly impoverished. It is impossible to command a wage where no one is able to pay the price. If wives work for husbands then their employers would be entitled to regulate their labor, to control standards, to penalize for poor work, and to replace incompetent and recalcitrant employees. Such employees would have to be charged for a number of household services rendered to them by their husband-employers. Now clearly, there exists no sociolegal framework for these arrangements. Nor can its nature be presumed upon, anymore than a legal framework was easily or universally implemented in the industrial world from which it is to be taken. Are all households to be inspected like factories? Do they form a competitive market, so that some households might have a DOW rating? Or is a civil service model the appropriate one? If these arrangements are too costly, then husband-employers cannot afford them, and presumably wives will be laid off or given unemployment pay. What about quality control in housework, sex, and mothering? If these are to be considered careers (rather than jobs) are they going to be regulated in accordance with universalistic achievement and norms involving the incumbents in a national and international field of competitive achievement? Merely to ask such questions should be enough to make it evident that "working women" are not a proper object of economic knowledge. They are, of course, a proper object of *ideology*. The problem is to try to grasp how women's ideological status requires the discursive production of feminist economics and its accompanying legislation. Thus the ERA in effect legislates defamilization inasmuch as it equalizes the familial roles of men and women, neutralizes assumptions about patrimony and alimony, or compassionate leave in the interests of maximizing conjugal participation in the labor force. In other words, the family is rendered absolutely flexible, its members interchangeable in order to accommodate the market requirements of free contractual agents. *The family is reduced to a contract to contract.* Thus from a Hobbesian perspective, the modern family merely shelters those who contract away their sexual powers over themselves in favor of a semisolipsistic orgasmic quota. From a Durkheimian prospective, the new family is separated from the precontractual fund that underwrites all secondary social contracts. Instead, the family is reconceived in contract so that its services are on an equal footing with any other contractual conduct of its members. Thus, in effect, *we are now legal members of a market society before we are members of a family.*

SOME ISSUES IN FEMINIST PSYCHOANALYSIS

We now see that the ideological function of juridico-legal bio-rights and the supremacy of legal contract is to underwrite the *state production of society.* But this requires defamilized individuals legally protected against their families of origin. Thus, as we shall see later on, the state must acquire further interests in the bio-reproductive environment. Curiously enough, the state's interests are serviced by *feminist psychoanalysis* which is the complement of feminist economics and law. Here feminist ideology is grounded in the symbolic property of the clitoris and a consummatory ethic of orgasm. This ideology is in turn underwritten by scientific and reductive physiological research. The solipsistic vision of modern sexology is, of course, defamiliz-ing. At the same time, it binds the individual directly to the therapeutic state which fosters its production in defense of the privatization of the juridico-legal subject:

> In every respect, then, Masters and Johnson's work represents the most extravagant assertion of the modernist theory of masturbation. Pursued to its logical conclusion that theory envisions a radically atomistic sex-ual order, in which the individual would achieve the same autonomy in his sexual life as he has come to expect in his religious and political life. Sex, one might say, is the last and most intimate form of dependen-cy, and Masters and Johnson would appear determined to liberate us from it. Thus it is not too farfetched to view them as the intellectual descendants of John Stuart Mill and Ralph Waldo Emerson. Self-abuse, they seem to be telling us, is the ultimate form of self-reliance.[7]

Sexual solipsism derives from feminist psychoanalysis. Freud, we are told, is responsible for a phallocentric world-view which is in fact motivated by male sexual inadequacy. Freudian woman derives from a conspiratorial sub-ordination of potentially insatiable clitoral sexuality to vaginal coitus initiat-ed by the dominant but anxious male. The result is allegedly a female cul-ture riddled with passivity, jealousy, penis-envy, and a poor sense of social justice and trade unionism. This is, of course, a travesty of Freud's views.[8] As I see it, the oedipal complex represents a retrospective myth whereby we recapture the shift from nature to society. The elementary unit of society is not the biological family. Rather, the facts of the biological family only become identifiable through the structural relation, or kinship, between fam-ilies. Thus the phallus (and not the penis, nor the vagina, for that matter) represents the symbolic exchange of brothers-and-sisters-in-law for sons and daughters. Thus husbands and wives acquire sons-and-daughters-in-law through the exchange of the children in exogamous marriage—and not by incest. In this system no one can *have* the phallus nor can anyone "own" their body. The former merely symbolizes a lawful exchange between males and females whose respective genitals are the instrument of the bio-psychic connection whereby they reproduce themselves with a family name—and not in the dark night of promiscuity. Thus the family represents what Vico calls a severe poem, the very origin of our humanity.

Perhaps the greatest impediment of feminist ideology is that it is likely to reduce our understanding of the forces that shape men and women's lives in relation to one another and to their children by alleging *sexism* to be a cause whereas sexism is itself shaped by the broad culture of power, competition, class, and colonialism in which the majority of men suffer along with women and children. Certainly, nothing suits the obfuscation of power better than the substitution of the battle of the sexes in place of the class and colonial struggle.

> The most intractable of the factors breeding sexual distortion are likely to be those that reach beyond the gender system, and which may be even more embedded in our culture. These include the general tendency toward depersonalization in American life, and our characteristic preoccupation with techniques and results. None of these tendencies is specifically sexual. Yet . . . they exert strong influence on our thinking about sex and our sexual behaviors. The same is true of America's pervasive socio-economic inequality. The individualistic-competitive ethos engendered by our particular brand of capitalism seriously hinders the achievement of sexual decency. As competition for limited, and narrowly defined, sexual goals becomes ever more ruthless, sexual unhappiness and victimization are likely to result. When this factor is placed alongside the pervasive inequality that fosters the commercial market in sex, we can see that economic considerations significantly shape the social patterning of sex in America.[9]

FROM FAMILY WAGE TO SOCIAL WAGE

When we adopt a structural approach to the economy of reproduction, as we have been doing, a number of things ought to stand out more clearly. First of all, economic reproduction is a *social* phenomenon that is broader than the idea of an *individual* wage. Individually and collectively, social and economic reproduction is a family phenomenon; it is motivated by more than a concern for consumption. We are family beings before we are economic beings. Hence the *family wage* is the basic economic unit of income. Young boys, girls, married women are therefore not legitimate recipients of family wages even when doing the "same" work. Unmarried mothers are even more of an anomaly in the family wage system—as would be gay family incomes. The family wage is paid in exchange for the work of being a familied/familizing employee. What is being recognized is a social function as well as a narrowly economic function. Women are therefore mistaken in thinking that it is men who are paid wages, or that the family wage is a man's property. Or even that half the man's wage is hers, since a family does not have two halves (though a marriage does.) They therefore will find anomalies in their experience with women's wages.

We should now look more closely at how it is women escape from "unpaid labor" in the home to paid labor in the market place. Here we shall only point to the input/output cycle in women's work. It needs to be remembered that the price/wage cycle leaves totally undescribed the actual social process of work-for-a-wage and the work-of-consumption. Here we are referring to the *feminization* of large areas of the working force and its mirror-image in the feminization of buying and selling.[10] Women then, can only withdraw their contribution to the various levels of the household economy, to the extent that they can either replace themselves with substitutes or surrogates, or else complement their own residual efforts. Thus women's work in the market place is tied to women's consumption in the market in the following ways:

1. Married women are relieved of some of the burden of domestic labor and so have time to do paid work, but simultaneously become *more dependent* on commodities, e.g. food, clothes, by going out to work.
2. Also, in part, they go out to work for these commodities, i.e., labor saving devices which are just outside the price range of a single wage. (This is not to suggest that most women do not do paid work to provide subsistence necessities).[11]
3. They work in those industries manufacturing such commodities: assembling electrical equipment, in light engineering, as well as in food processing.
4. For some young, single women—particularly those employed as boutique assistants, receptionists, air hostesses—consumption of certain commodities, i.e., make-up and fashion (in which trades it is again women who are primarily the employees), is essential to the sale of their labor power.
5. This same group plus other women work in the distribution/marketing apparatus which promotes and sells such commodities.[12]

Just as this cycle of working women exists in the market place, a similar cycle exists for women in the government sector. If the family wage is supplemented/exchanged for women's work in the market, so too the social wage is supplemented/exchanged for women's work in

6. The health services industries where women are the main employees;
7. School, day care and similar agencies;
8. The governmental bureaucracies that administer themselves and the health, education and welfare institutions.

It is as a wife and a mother that the woman has a claim (however it is allocated) upon the family income. But in no sense is her claim a wage for her labor comparable to the family wage received against the social reproduction of the familied labor force of males. If men are interested in a "living wage" it is because they are held responsible for the social reproduction of their labor through the family. Anyone familiar with the working man's pay check will know that what is left of it for his individual use is next to nothing. Indeed, he may well have kept more for himself from his wages as a boy. It is therefore an illusion for women to imagine that their wages will reward them much more richly. Of course, this is a promising prospect where women pursue professional incomes whose residual spending is much higher than what wages afford. It is even more so where women hope to double the professional family income and the style of consumption it offers. But most women, like most men, will work for wages and either to supplement or gain solely the family wage.

The idea of *individual wages* is therefore quite problematic. Even in the narrowly productive sphere it would be difficult to give it a precise definition. The notion is rather a *social fiction* resonant with Lockean ideals of the right to one's own property but in fact belied by the collective nature of work, technology, and science which make individual effort productive. Much of feminist economics is ideologically tied to the Lockean prospect of possessive individualism, if ever women could stake out a claim to their own bodies before society sets in.[13] The real context of the family wage, however, is a world in which most men and women do not own the means of production and only exchange their labor against a family wage that reproduces the socioeconomic system.[14]

Now historically capitalism has not been generous with labor. Yet it has its limits. Slavery is not a formal option, just as complete automation, although formally a strategy, is also ruled out. Early capitalism, therefore, had to adjust itself to familied labor. That is to say, it was necessary to pass factory, school, health, and municipal legislation which allowed the social reproduction of capitalism to avoid self-destructive cruelty, communality, ignorance, and disease. Of course, it did so within its limitations as a class system—the working-class family wage being far less in sum and security than the income of other classes. And, as is well-known, orphans, widows, spinsters, the sick and old men without families, unless they found themselves in the upper classes fell to the bottom of society. In such a world, being inside the family was incomparably better than trying to survive outside it. However, it will not do to argue that the family is merely an instrument of capitalist reproduction. This is to reverse the order of things. The experience of early capitalism made it clear that there are limits to the convenience of barbarizing society by impoverishing and criminalizing its lower orders at the same time requiring of them safe and moral conduct.

But since the dynamics of capitalism are such that it increasingly socializes itself rather than be communized, we find that in late capitalism there is a large amount of legislation—not always coherent— which we shall interpret as moving *from the family wage to the social wage*[15] as the basic social accounting unit. It should be remembered, however, that we have already argued that vis-á-vis the notion of individual income, all income is social, the family wage being the basic element in what we shall now call the *social wage*, including various transfer payments of a social welfare nature. The social wage is a composite notion of what society generally conceives to be a minimal standard of living and what various classes, occupations and professions consider society owes to them. "Society" is understood to mean what welfare state capitalism implies for individuals, families, and classes in its continued reproduction as a political and economic system.

THE FEMINIZATION OF SOCIOLEGAL DISCOURSE IN LATE CAPITALISM

The evolution of the social wage involves a recent legislative history around (a) full employment, (b) social insurance, (c) social medicine, and (d) universal education.

Obviously, we are abstracting from an extremely complex political and legislative history. The risk in this is only justified by the potential insight into a discernible direction that we had in mind by speaking of the *feminization of law under late capitalism*. Here, again, we have a concept that is not perfectly congruent with the phenomena it seeks to elicit. Rather, it emphasizes a single mode of disclosure in what is actually an area of multiple, contradictory discourses that (a) support the family, (b) regulate the female body, and (c) supercede (defamilize) the family.

Among legislation supportive of the family we think of family allowances, health care, day care, school milk, social work, family and juvenile courts. Other legislation may be regarded as supportive or as defamilizing, depending upon how one interprets legislation that we shall treat as regulating the female body—contraception, abortion, rape, divorce, separation. Thus legislation concerning prostitution and homosexuality either protects or defamilizes the family (to some extent both). Or it is to be seen as part of the state legislation of purely contractual sexuality which in turn requires much of the legislation regulating women's bodies.

Thus the law is embarked upon a double discourse of *feminization and defamilization of the embodied legal subject* as the ideological agent of all private and social contracts. Women experience this double tendency towards feminization and defamilization by and large by taking on the stress of combining opposing roles and ideologies of family and individual work. Social scientists discover new trends in family styles, sex roles, abortion, and androgynous psychology which are said to represent the next stage of human development. What is more likely is that these are adjustments to the production and consumption requirements of late capitalism in which law and the social sciences discover the very society their own practice presupposes.

In speaking of "late capitalism" we have in mind an extreme tendency of early capitalism, rather than any theory of current economic crisis. What is the tendency to which we refer the continuous efforts of capitalism? Despite contrary opinion, we think it is not exploitation, class war, profitability, or human degradation. We do not deny these phenomena. Rather, we think they are embedded in an *extreme tendency of capitalism*, apparent from the early factory system, *to replace human beings with machines*. Thus the factory system from the very beginning pitted men against women and children, that is, it divided the family against itself:

> Along with the tool, the skill of the workman in handling it passes over to the machine. The capabilities of the tool are emancipated from the restraints that are inseparable from human labour-power. Thereby the technical foundation on which is based the division of labour in manufacture, is swept away. Hence, in the place of the hierarchy of specialized workmen that characterizes manufacture, there steps, in *the automatic factory*, a tendency to equalize and to reduce to one and the same level every kind of work that has to be done by the minders of the machines; in the place of the artificially produced differentiations of the detail workmen, step natural differences of age and sex.[16]

To the extent that capitalism still requires human labor, therefore, it is engaged in a double tendency at once to require the highest level of skilled machine minding and to employ the least skilled labor of women and children. This double tendency is at its extreme in late capitalism where it is possible to exploit the relative levels of skilled and unskilled labor in the world market. Thus (from the cotton industry to the computer chip) Third-World women and children today replace the women and children of 18th- and 19th-century Europe:

> In so far as machinery dispenses with muscular power, it becomes a means of employing labourers of slight muscular strength, and those whose bodily development is incomplete, but whose limbs are all the

more supple. The labour of women and children was, therefore, the first thing sought for by capitalists who used machinery. That mighty substitute for labour and labourers was forthwith changed into a means for increasing the number of wage-labourers by enrolling, under the direct sway of capital, every member of the workman's family, without distinction of age or sex.[17]

Thus late capitalism operates with a *dual economy*[18] of labor in which women are likely to find themselves alongside all male, immigrant, ethnic, and colonial workers whose low-skill, nonunion, temporary and other characteristics required by capitalism's extreme tendency to replace human beings with machines.[19] One might think that in this line capitalism would not balk at employing even more degraded labor, such as criminals, the insane, severely handicapped people, and proverbial monkeys. So far these possibilities have remained science fictional and will probably do so in the future. This is because the lord of creation sees himself in *imago machinae*, even when it replaces him, but is unwilling to see himself in a degraded human being. By and large, capitalists share this morality, while scientists risk madness in denying it. Of course, capitalism encourages scientists to produce men—and in biotechnology we are witnessing a new frontier in this fiction. But in practice capitalism settles for its own commercial and state production of society, and thus it has a special concern for defamilism and feminism.

It must be remembered that early industrial capitalism did not immediately destroy the domestic economy. Even into the early 20th century the family economy supplemented the industrial wage. Gradually, of course, the increasing density of urbanization and industrialization eroded the family economy, moving more and more of its production and consumption into the commodity form and female employment.[20] Food, clothing, entertainment became market activities purchased outside the household. The latter became less and less the center of individual and community life. As a result, the incidence of crime, mental illness and old age became the responsibility of public service and professional agencies. The latter, like the economy they serve, engaged in a double discourse of defamilism and feminism.

With this shift, the deepest tendency of capitalism begins to reveal itself: *because capitalism seeks, but never completely manages, to replace human beings with machines, it is driven to replace family society with a consumer and service society underwritten by its industrial, legal, medical technology, and feminizing ideologies.* What I have in mind here is not just a simple progressive accumulation of goods and services by the household of early capitalism. What is involved is a sociolegal redefinition of the family for the consumption tasks of late capitalism. A major pedagogic and thera-peutic switch is involved in which family attitudes are "engineered" on behalf of the industrial, commercial, and state system of late capitalism. The family is increasingly subjected to a huge degradation in favor of the family wise in consumption. This involves the degradation of household work, cooking, cleaning, caring in any way that does not bring the family into the orbit of industry, commerce, and professionalism. Simultaneously, the fami-ly that is wired into such commercialism and professionalism in the delivery of its functions is elevated. This means, of course, that the family again splits. *The bourgeois family whose professional members service themselves in saving the working-class family, become the principal circuit of defamilized and feminized discourse upon family health, education, and welfare.* Thus children's health, education, and consumer awareness are the discursive channels for the reorientation of the family to the new demands of late capitalism. Bourgeois feminism, the legal, medical, and educational professions, as well as commercial advertising, combine to subordinate the family "patriarch" to his more enlightened women and children:

> In the death of patriarchy, both libertarians and business shared an interest. Yet their interests were at odds with one another. . . . The com-modified answers to the question of "how to live" began to take on a distinctive character. Utilizing the collective image of the family, the ads in their contribution to mass culture did their best to deny that col-lectivity. Each aspect of the family *collective*—the source of decision making, the locus of child rearing, the things which elicited affectionate response—all of these now pointed outward toward the world of com-modities for their direction. Corporate America had begun to define itself as the father of us all.[21]

It is an easy task to recall endless advertisements, comic strips, car-toons, kids' movies, and family movies that dramatize the end of the patriar-chal family and its surrender to *consumer matriarchy* sponsored by pseudo-paternalist corporations. Much of this is what has made Hollywood America the symbol of freedom for millions of people whose families, marriages, and communities kept their noses to the grindstone of scarcity and authority. It is the staple of American lawlessness smiled upon by American law, itself so often the clown in the comedy. In reality, the American family is as much exploited in these scenarios as is the still grim industrial and urban settings that furnish their background. The tragicomedy, however, is played out dif-ferently in the towns and in the country, in the middle and upper classes and in the working class, immigrant families. There arise huge problems of pub-lic health and morality, of criminality and ignorance which beset the ideology of individual self-development. These, however, provide for a double response, at once reinforcing individualism while simultaneously professionalizing and bureaucratizing the sociolegal practices that correct for its failures.[22] Again, in this the various discourses of scientism, individualism, defamilism, and feminism are subtly interwoven.

How strong this pattern of sociolegal discourse is, may be seen if we turn from the Anglo-American context to France where we have similar relations between the state, medicine, feminism, and defamilism in the bourgeois and working-class domains. Once again, the differential impact of industrialism and urbanism upon the family, public health and safety, as well upon the potential for socioeconomic reproduction, forms the basis for a discursive alliance between the state and the liberal professions. Specific issues like working-class housing, female and child labor, public hygiene, criminality, and the moral fabric of society were launched in the name of the collective family. In practice, it involved a double strategy with respect to working-class and bourgeois families which we might call:

1. The *confinement* of the working class family: "The problem was to organize a space large enough to be hygienic, small enough so that only the family could live in it, and distributed in such a way that the parents could monitor their children. Housing had to become a factor that complemented the school in the supervision of children: its mobile elements had to be excluded so that children might be immobilized within it. The search for intimacy and the domestic jurisdiction that was proposed to the working class woman were the means to make this dwelling acceptable, and even attractive, in the transition from a schema that was tied to production and social life to a conception based on separation and surveillance."[23]
2. The *feminization* of the bourgeois family: "Owing to the revalorization of educative tasks, a new continuity was established for the bourgeois woman between her family activities and her social activities. She discovered a new missionary domain in which to operate; a new professional sphere was opened, consisting in the spread of the new welfare and educational norms."[24]

Once again, we have to see that the Marxist and feminist critique of the family as a reproductive committee of the bourgeoisie and the state will not do.[25] The analytic task is to see how divergent discursive strategies developed around the family as a bulwark against the state and simultaneously as the limit of criticism and revolt directed against the social order. In other words, the liberal bourgeois conception of state and economic relations meant that the bourgeoisie had to find a solution to the problem of pauperism without generating socialism even though granting rights to work, education, and welfare. Simultaneously, the bourgeoisie had to find a new basis for social commitment on the part of the masses while excluding them from political participation. As Donzelot shows, there was a preference for two strategies of control: (a) *Philanthropy*, and (b) *Medical-Hygiene*. The two strategies were designed to transform the family into a buffer between pauperism on the one hand, by shoring up the practice of family savings and family assistance, and from irresponsible patriarchy, by defending standards of health and morality due to children. Thus from both sides the family became the focus of philanthropic and therapeutic strategies designed to raise its reproductive potential with respect to the economy and the social order, short of absolute statist strategies. By the same token, the family was saved by reducing its autonomy vis-á-vis the therapeutic state which served the liberal bourgeois concept of society without socialism. It is no accident, then, that the laws on divorce go hand in hand with state laws that undermine patriarchal and familial authority over children: the liberality of the marriage contract is exchanged for the state as parent in the last instance:

> the modern family is not so much an institution as a *mechanism*. It is through the disparity of the familial configurations (the working class and bourgeois bipolarity), the variances between individual interests and the family interest, that this mechanism operates. Its strength lies in a social *architectonics* whose characteristic feature is always to couple an exterior intervention with conflicts or differences of potential within the family: the protection of poor children which allowed for the destruction of the family as an island of resistance; the privileged alliance of the doctor and the educator with the wife for developing procedures of savings, educational promotion and so on. The procedures of social control depend much more on the complexity of intrafamilial relationships than on its complexes, more on its craving for betterment than on the defense of its acquisitions (private property, judicial rigidity). A wonderful mechanism, since it enables the social body to deal with marginality through a near total dispossession of private rights, and to encourage positive integration, the renunciation of the question of political right through the private pursuit of well-being[26]

We have always to remember that the tendencies we are describing are never in practice wholly congruent. Thus it is possible to see much family law as having delivered married women from the authority of their husbands, restoring child custody and releasing them from sexual monogamy.[27] These changes have considerably altered the intrafamilial status of wives.

However, the socioeconomic reality observed by the law is that child care is best exercised within the family household and that the public costs where this fails is to be shifted to individuals. Since women continue to be weak wage earners, while the state holds out on day care service, women's rights in family law terms do not match the structural realities of the working women's economy. It is therefore a difficult matter to decide upon the extent to which the state oppresses women rather than men. Indeed, the question has no general answer in this form. It can only be approached relative to specific historical stages and policies of capitalism and the liberal welfare state. The ideological function of welfare, social work, and social policy as state apparatuses has been noticed by Wilson. This does not imply any conspiracy on the part of the state or of the bourgeoisie. Capitalist society consists of competing and incongruent interests whose group and class affiliations are modified by the ideology of welfare. The position of women, children, and the nuclear family varies with respect to these conflicting claims. As I have argued here, just as the child in the factory and workhouse was the concern of early Victorian social legislation, the position of the child in the nuclear family is the focus of modern welfare legislation, despite all kinds of changes in familial relations. The implications for women cannot be considered apart from this constellation:

> social policy is simply one aspect of the capitalist state, an acceptable face of capitalism, and social welfare policies amount to no less than the *state organization of domestic life*. Women encounter state repression within the very bosom of the family. This may seem paradoxical when the ideology of individualism and private property that has grown with capitalism has stressed the sanctity of family privacy. But in many ways the Welfare State, like the position of women, is full of paradox and contradiction.[28]

Thus the welfare state makes a greater effort through social work and psychiatric practice to keep child care in the home, whereas it is more determined to move health services such as pre- and postnatal care, birthing, birth control, and abortion out of the home, along with old age and death. Ultimately, however, the tendency of the therapeutic state is to increase its power over the *body politic*. Here it would be necessary to show how new biorights to body parts, transfusions, transplants and the futuristic possibilities of biotechnological engineering constitute the latest contexts of medicolegal discourse and state power.

CONCLUSION

As I see it, the shift from early to late capitalism has involved a double strategy of socio-legal redefinition of familism and feminism involving the emergence of a new *biopolitical economy* suited to the needs of liberal welfare state capitalism. This shift may be roughly summarized in terms of the following stages:

1. *Preindustrial family economy*: farm and handicraft work employing the whole family.
2. *Early capitalism*
   (a) family is moved into the factory;
   (b) the family is legislated out of the factory, and
   In the bourgeois family
   (c) women are feminized; and
   (d) their children are moralized, while
   In the working class
   (e) women are feminized but work; and
   (f) their children are moralized, receive some schooling and soon work.
3. *Late Capitalism*
   (a) the factory is moved into the family, i.e., consumption provides models for work;
   (b) consumption is feminized and infantilized;
   (c) women's bodies are feminized for work and consumption;
   (d) In both the bourgeois family and the working-class family, the legal and medical sciences are the common source of defamilism and feminism; and
   (e) The welfare state legislates transfer payments to augment the working-class family wage into a social wage; while
   (f) the social sciences provide a legal, administrative and therapeutic culture in which the twin discourses of defamilism and feminism are floated for all classes.

FEMINIZED CAPITALISM

Today capitalism advances the emancipation of women as producers and consumers by according to them the right to control their own bodies—aborting not only the foetus but all social relationships that limit the myth of modernity. Thus women's bodies become the site of the political struggle between the right and the left in all industrialized societies where *genderless individualism* is the norm:

> An industrial society cannot exist unless it imposes certain unisex assumptions: the assumptions that both sexes are made for the same work, perceive the same reality, and have, with some minor cosmetic variations, the same needs. And the assumption of scarcity, which is fundamental to economics, is itself logically based on this unisex postulate.[29]

Of course, much now depends upon how global capitalism and deficit reduction will restructure welfare states in relation to the needs of the new working family.

NOTES

1.   I am aware that recent historical scholarship has reworked many of the underlying assumptions in the way we schematize the position of the family in early and late capitalism. See Alan Macfarlane, *The Culture of Capitalism* (Oxford: Basil Blackwell, 1987); also, *The Family in Political Thought.* Edited by Jean Bethke Elshtain (Amherst: University of Massachusetts Press, 1982).

2.   Claude Levi-Strauss, "Language and the Analysis of Social Laws," in his *Structural Anthropology*, trans. by Claire Jacobson and Brooke Grundfest Schoepf (New York: Basic Books, 1963), p. 61. For commentary see Pierre Maranda, "Structuralism in Cultural Anthropology," *Annual Review of Anthropology*, Vol. I (1972):329-348; Marshall D. Sahlins, "On the Sociology of Primitive Exchange," in *The Relevance of Models for Social Anthropology*, Michael Banton, ed. (London: Tavistock Publications, 1965), pp. 139-236.

3.   Juliet Mitchell, "Patriarchy, Kinship and Women as Exchange Objects," in her *Psychoanalysis and Feminism* (Harmondsworth: Penguin Books, 1975), pp. 370-376.

4.   Philip Rieff, *The Triumph of the Therapeutic: Uses of Faith after Freud* (London: Chatto and Windus, 1966).

5.   John O'Neill, *Five Bodies: The Human Shape of Modern Society* (Ithaca: Cornell University Press, 1985); *The Missing Child in Liberal Theory* (1994).

6.   Karl Marx, *Capital: A Critique of Political Economy* (New York: The Modern Library, 1906, original Charles H. Kerr), p. 395.

7.   Paul A. Robinson, *The Modernization of Sex: Havelock Ellis, Alfred Kinsey, William Masters* (New York: Harper & Row, 1976), p. 144.

8.   For a corrective see Juliet Mitchell, *Psychoanalysis and Feminism*.

9.   Edwin M. Schur, *The Americanization of Sex* (Philadelphia: Temple University Press, 1988), p. 199.

10.   On the *class* significance of the feminization process, see the discussion of the theories of Giddens, Braverman and Poulantzas by Jackie West, "Women, Sex and Class," in *Materialism, Women and Modes of Production*, Annette Kuhn and AnnMarie Wolpe, eds. (London: Routledge and Kegan Paul, 1978), pp. 220-253. For the *colonial*, or third world significance of feminization, see Michele Mattelhart, *Women, Media and Crisis: Femininity and Disorder* (London: Comedia Publishing Group, 1986).

11.   Ruth Schwartz Cowan, *More Work for Mother: The Ironies of Household Technology from the Open Hearth to the Microwave* (New York: Basic Books, 1983).

12.   Lucy Bland, ed., "Women 'Inside and Outside' the relations of production," in *Women Take Issue: Aspects of Women's Subordination* (London: Hutchinson, 1978), pp. 35-78; also *Feminization of the Labor Force: Paradoxes and Promises*, eds. Jane Jenson, Elisabeth Hagen and Ceallaigh Reddy (New York: Oxford University Press, 1981).

13.   Alison Jagger, "Political Philosophies of Women's Liberation," in *Feminism and Philosophy*, Mary Vetterling-Braggin, Frederick A. Elliston and Jane English, eds. (Totowa, NJ: Rowman & Littlefield, 1977), pp. 5-21; Carole Pateman, *The Sexual Contract* (Oxford: Basil Blackwell, 1988).

14.   Victor R. Fuchs, *Women's Quest for Economic Equality* (Cambridge: Harvard University Press, 1988).

15.   Paul Adams, "Social Control of Social Wage: On the Political Economy of the Welfare State," in *Education and the State, Volume II: Politics, Patriarchy and Practice*. Roger Dale, ed. (Lewes, England: Falmer Press, 1981), pp. 231-238.

16.   Marx, *Capital*, p. 420.

17.   *Ibid.*, pp. 489-490.

18.   R. D. Barron and G.M. Norris, "Sexual Divisions and the Dual Labour Market," in *Dependence and Exploitation in Work and Marriage*, Diana Leonard Barker and Sheila Allen, eds. (London: Longman, 1976), pp. 47-69.

19.   Veronica Beechey, "Women and Production: A Critical Analysis of Some Sociological Theories of Women's Work," in *Education and the State, Volume II*, pp. 115-140.

20.   Harry Braverman, *Labor and Monopoly Capital: The Degradation of Work in the Twentieth Century* (New York: Monthly Review Press, 1974).

21.   Stuart Ewen, *Captains of Consciousness: Advertising and the Social Roots of the Consumer Culture* (New York: McGraw-Hill, 1976), p. 184.

22.   Burton J. Bledstein, *The Culture of Professionalism: The Middle Class and the Development of Higher Education in America* (New York: W.W. Norton, 1978).

23.   Jacques Donzelot, *The Policing of Families* (New York: Pantheon Books, 1979), p. 44.

24.   *Ibid.*, pp. 45-46.

25. *The Unhappy Marriage of Marxism and Feminism: A Debate on Class and Patriarchy*, Lydia Sargent, ed. (London: Phi Press), 1986.

26. *The Policing of Families*, p. 94.

27. Julia Brophy and Carol Smart, " From Disregard to Disrepute: The Position of Women in Family Law, *Feminist Review* 9 (1981):3-15; Mary McIntosh, "The State and the Oppression of Women," in *Feminism and Materialism*, pp. 254-289.

28. Elizabeth Wilson, *Women and the Welfare State* (London: Tavistock, 1977).

29. Ivan Illich, *Gender* (New York: Pantheon Books, 1982), pp. 9-10; H.T. Wilson, *Sex and Gender: Making Cultural Sense of Civilization* (Leiden: E.J. Brill, 1989).

CHAPTER FIVE

Political Communication: The Corporate Agenda and the Legitimation Problem

There are numerous ways to approach the legitimation of power and communication in industrial societies. *We shall treat the legitimation process as a communicative task addressed to the mobilization of members' commitment to the goals and institutionalized allocations of resources that translate social goals into daily conveniences, rewards, and punishments.*[1] Along these lines we can formulate a first gloss on the legitimation problem in the following terms: the *legitimacy* of any political system requires that (a) its *members* have access to the channels whereby social *goals* are articulated, (b) to varying degrees, members are aware of and feel entitled to exercise their *rights* in the translation of subjective *needs* into specific and local allocations of institutional *resources*, and (c) members' troubles with the determinate processes of resource allocation establish prima facie claims for *reforms* at level (b) or *change* of goals at level (a).

The interrelationship or loop-effect between the preceding processes represents a *normative* formulation of the legitimation process as a *communicative community*. Whenever there are gross faults in the articulation of goals, access to means, deprivation of rights, in short, where there are practices of misinformation, false consciousness, and ignorance, as well as forceful exclusions, we may speak of *repressive communication*.[2] In strictly normative terms, the practices of repressive communication lower the *legitimacy* of the political system. In practice, however, they may well contribute to its integration as a system of *power*.

The preceding propositions stand as a gloss or *idealization* upon the communicative processes they seem to describe (hence the necessity of the material in the surrounding chapters) inasmuch as they are perfectly elliptical with respect to the endogenous orders of practical reasoning, expressions, and displays that accomplish the everyday work of legitimation. They provide a *reasoned conjecture* with respect to the problem in hand whose *sensible nature* otherwise remains for study as the situated and endogenous communications through which members provide for the work of political legitimation.[3]

In view of the intractability of the separation between ethical and repressive communication that may nevertheless integrate the political system, we are obliged to reconjecture the legitimation problem as the communicative task of maximizing the *ethical probability* that the political system will mobilize commitment, apart from the consideration of members' actual beliefs and loyalties, and in ways that are recognizable resources of rule and participation:

> Naturally, the legitimacy of a system of domination may be treated sociologically only as the probability that to a relevant degree the appropriate attitudes will exist, and the corresponding political conduct ensue. . . . What is important is the fact that in a given case the particular claim to legitimacy is to a significant degree, and according to its type, treated as "valid"; that this fact confirms the position of the persons claiming authority and that it helps to determine the choice of means of its exercise.[4]

In the simplest terms, the legitimacy of a political system depends upon its ability to deliver the goods of the good life and to have its members believe in its very ability to do so, without more than occasional resort to *force*, *strikes*, and *civil violence*. Thus, in modern industrial society, the legitimacy of the political system is highly dependent upon its members' sense of participation in the distribution of economic, health, educational, and other welfare goods. In capitalist society, the legitimation problem is the matter of how it is that capitalist society continues to distribute its goods unequally and yet legitimately, at least in so far as members are willing to live with its inconveniences and injustice, short of rebellion (see Chapter Ten).

We shall assume that ultimately the "validity" of the capitalist political system rests in the reference of its goals and resource allocations to parliamentary and electoral ratification which bring political decisions within the *social contract*. In other words, it is characteristic of the processes of political legitimation in capitalist society that together with parliament they involve public debate, public information, news media, journalism, and the institutions of public education. These institutions are conceived in the main to realize the belief that political life can be exemplary of rational and just conduct at the same time that the polity accommodates the sheer necessities and incommodities of economic and social life. The process of legitimation and delegitimation, however, assume particular features in the present context of state-administered *capitalism*.[5] It is essential to the understanding of this context that we grasp the legitimation process as an *intrinsic* function of political economy and not an epiphenomenal exercise, as Marxists might argue. Very simply, a large part of the present capitalist economy consists in the activity of state enterprises which do not merely complement or correct the private-market system but largely determine the parameters of economic activity. Therefore, it is an essential task of the legitimation process to secure allegiance to the new schedule of public and private production.[6] A second major feature of this process is the political socialization of large numbers of persons who enjoy goods and services without direct employment in either sector of the economy. In short, employment in the public sector and consumption of health, education, and welfare from that sector without corresponding inputs defines a large population whose loyalty to the political system cannot be identified in terms of previous class and motivational attitudes.

Under these conditions, the legitimation problem acquires the following distinct features:

1. The market system can no longer be legitimated in terms of its power to mobilize the bourgeois ideology of private exchange;
2. The central administrative role of the state in the economy must be legitimated;
3. The administrative role of the state must be maintained independently of the mechanisms of formal participation in democratic politics;
4. The legitimation process therefore requires the mobilization of diffuse mass loyalty without direct participation;
5. The depolitization of the public realm requires (a) civil privatism, i.e., the pursuit of consumption, leisure, and careers in exchange for political abstinence, and (b) the ideological justification of public depoliticization by means of elitist theories of the democ-

ratic process or by technocratic and professionalized accounting practices which rationalize administrative power.[7]

The success of corporate capitalism in controlling its environment has won for it political acceptance from its employees and the stabilizing support of state-administered antidepressants for those moments in which the soulfulness of the corporation threatens to reach a low point. The corporation's control of its environment assumes a variety of forms, ranging from its ability to control price-cost relationships, levels and composition of investment, the nature of research and innovation, the location of industry with its effects upon local governmental intervention and, last but not least, the power to shape the physical and sociopsychological environment of the consumer public. In each case, these powers of the corporation are of enormous social and political consequence.[8]

The corporate practice of integrating its production and sales efforts through the generation of wants, thereby engineering consumer response provides an *ex post facto* ratification of its commitment of social resources as determined by the corporate agenda. While paying lip service to consumer sovereignty in the final allocation of social resources, the corporation can in fact assume the conventional distribution of social resources between the public and private sectors of the economy. This presumption is a political reality inasmuch as the demand for public services presently arises out of the needs of low-income groups who are powerless to compete away social resources from the private uses of higher-income groups. It is only in the context of the unequal distribution of income, which remains as much as ever a defining characteristic of affluent capitalism, that one can properly understand the imperative of production or, rather, of *relative overproduction for the private sector*, which in turn promotes the secondary imperatives of consumption and other direction. Despite the heralds of the age of high mass consumption, the fact is that monopoly capitalism is a production system continually faced with underconsumption.

For want of a genuinely public political economy where the political and social activities of individuals can achieve a focus and historical perspective, men abandon politics for the civic affairs of suburbia or the "bread and butter" questions of unionism. By shifting awareness toward improvements in consumption styles, these tactics deflect attention from the social imbalance which results from the pursuit of intraclass benefits that leave whole sectors of the population outside of their calculus. This tactic is further strengthened by the ideological acceptance of social improvement through the escalation effect of an expanding economy upon all classes rather than through any radical redistribution of class income or the extension of chances of individual mobility between classes.

As individual awareness is increasingly shifted toward a concern with consumption, economic knowledge is reduced to a concern with prices in abstraction from the corporate agenda which determines prices. The result is a loss of any coherent ideological awareness of the political and economic contexts of individual action. However, this does not represent an end of ideology. It is simply the nature of the dominant ideology of individualism shaped by the context of corporate capitalism. In order to break the tendency to monetize all individual experience, and in order to shift individual time perspectives away from short-term consumer expectations, it is necessary to institutionalize more universal goals of collective and long-term value. Such a requirement falls outside the pattern of instant satisfactions projected by the consumer orientation. Any concern with social balance, institutional poverty, and waste, or the interaction between politics, economics, and culture presupposes a collective and historical framework; but this is foreign to the liberal ideology of individual agency and its moralistic acceptance of inequality and failure within a natural order of social competition and private success.

With these considerations in mind, we now turn to the large issue of the role of communications in the polity. The legitimation process is characterized at all levels of politics, society, and economy by what we call a relative *linguistic privacy* which circumvents the ideals and practice of the communicative community. Several factors are at work in this process:

1. The rationalization of the administered society requires that discourse be problem-specific and subject to decisionistic or calculative reasoning;
2. The very scientificity of the language and reportage of the social sciences contributes to the administrative effort to manage behavior and institutions according to maximum efficiency standards;
3. The ability by and large of the administered society to command allegiance in exchange for participation in goods and services, reduces political participation to the demand for 'information' and the residual right to withdraw loyalty in elections;
4. The combined effects of these processes (1-3) upon the communicative competence of citizens is that discourse about the ideal values of political, economic and social life is marginalized and alienated as talk lacking any rational, i.e., decisionistic grammar.

The events of the 1960s have passed so far as their political style is concerned. But the issues have not fallen by the way. Indeed, the new global trade order has intensified them, giving rise to an unexpected street politics in the past year. Owing to the structural considerations we have already raised, it will be difficult to side with the occasioned rationality of citizen outbursts with regard to urban, school, hospital, transportation, and the variety of mundane issues that test the quality of everyday life in the body politic. In these scenarios, political commentary will be divided between the realists and the moralists.[9] The realists will see in the political process an occasion for administrative expertise that is beyond the claims of traditional morality and popular intervention. This claim requires the absolute separation of politics and society—the depoliticization of the public realm accompanied by civic privatism fed by consumerism and careerism. It calls for the end of political speech in favor of officialese, for the adoption of efficiency as a legitimating as well as an organizing principle. The moralists will see in the reduction of the legitimation process to rules of systems efficiency the end of rational political discourse. They will regard the vision of the realists as an absurd project dependent upon the destruction of ordinary adult competencies with processes of reasoning and justification that are constitutive features of the human world—let alone that "old European human dignity" for which Habermas has entered a plea in his debate with Luhmann.

Aware of the pitfall of either realism or moralism in political affairs, I shall formulate street politics and green tactics in terms of the classical metaphor of the body politic.[10] I believe this permits us to articulate a complex structure of needs and wants around the institutions of *family*, *economic*, and *personal life* in terms of which we can then critically evaluate the processes of political legitimation and integration. Thus I have differentiated the body politic into three levels of values or goal structures and the corresponding institutional systems that potentially realize and integrate the individual and social production of these values.

The *organic body politic* represents a way of collecting the interest men have in their well-being, their bodily health and reproduction. The welfare of the family is iconic of the satisfaction of these demands.

BODY POLITICS	(i) organic (ii) productive } *Values*	(i) family (ii) economy } *Institutional Resources*
	(iii) libidinal	(iii) person

The *productive body politic* represents the complex organization of labor and intellect expended in the social reproduction of the body politic. Here, too, we speak of a healthy family. The *libidinal body politic* represents a level of desire that fulfills the order of personality in so far as it transcends the goods of family and economy. So long as we continue to be birthed and familied of one another, then the bodily, social, and libidinal orders of living will not be separable pursuits. By the same token, the body politic cannot be reduced to purely economistic satisfactions any more than to the dream of love's body:

1. A distinctive feature of the metaphor of the body politic is that it allows us to stand away from the systems, that is, machine, cybernetic, and organization metaphors that reduce the problem of political legitimacy to sheerly cognitivist sciences.

2. This shift in turn recovers the plain rationalities of everyday living, family survival, health, self-respect, love, and communion. Citizens are aware of the necessary interrelationships between their family, economic, and personal commitments. They judge the benefits of their labors in the productive sector of the body politic in terms of the returns to their familial and personal lives. They are willing to make tradeoffs between the demands of family life and the ambitions of their personal and libidinal life. In short, citizens have a fairly complex understanding of their corporate life which is not reducible to the single pattern of utilitarian or decisionistic reasoning that governs calculations in the productive sector.

3. By differentiating these three levels of the body politic, we further separate ourselves from naturalistic accounts of the legitimacy problem by introducing a *logic of ethical development* as the fundamental myth of political life. The three levels of family, economic, and personal level represent an historical-ethical development and also permit us to identify contradictions or constraints and regressions in the body politic. Thus, we can identify *alienation* as a complex phenomenon that affects not only the productive body but also the organic and libidinal body. Conversely, alienation is not solved merely by satisfying organic needs, nor by the smooth engineering of productive relations since these do not meet the demands of the libidinal body. By the same token, we cannot abstract the dreams of libidinal life from our commitments to familial and economic life.

4. Thus a critical theory of the legitimacy problems of the body politic is simultaneously a constitutive theory of social development and of citizens' recognition of the places in their lives where this development is blocked and even deteriorating.

5. Citizens' expression of their experience with the underlying logic of development that sustains political legitimacy will not be limited to official electoral conduct. It will include strikes, family breakdown, crime, protest, lampoons, neighborhood and street gatherings, music, song, poster and wall art.

6. A critical theory of political legitimacy does not discount the rationality of citizens' ordinary accounts of their political experience in terms of the vocabularies of family, work, and person.

Moreover, it does not presume upon either the found rationality or irrationality of such accounts.

We cannot pursue here the detail of Habermas's theory of communicative competence,[11] although the reader may turn to Chapter 10 for further considerations based upon it. What we must understand about it for now is its place with respect to the *imperativist*, that is, command and decisionistic structures of the economy and the polity. Habermas's argument is that the legitimation problem is not reducible to the imperatives of an administered political economy. Rather, the latter processes are to be embedded in the *communicative community* which is not driven by imperatives but by the norms of dialogue, equal and rational exchange of opinions, and ideas working towards the best available consensus with a minimum of constraint and distortion:

> The speech situation, which is determined by pure intersubjectivity, is an idealization. The mastery of dialogue constitutive universals is not synonymous to the capability of actually establishing the ideal speech situation. But communicative competence does mean the mastery of the means of construction necessary for the establishment of an ideal speech situation. No matter how the inter-subjectivity of mutual understanding may be deformed, the design of an ideal speech situation is necessarily implied with the structure of potential speech; for every speech, even that of intentional deception, is orientated towards the idea of truth. This idea can only be analyzed with regard to a consensus achieved in unrestraint and universal discourse. Insofar as we master the means for the construction of an ideal speech situation, we can conceive the ideas of truth, freedom and justice—which interpret each other—only as ideas of course. For on the strength of communicative competence we can by no means really produce the ideal speech situation independent of the empirical structures of the social system to which we belong; we can only anticipate this situation.[12]

In my own view, the politics of communicative competence cannot be settled in the reciprocal presuppositions of an *ideal-speech situation*. For unless the *pedagogic practices* between the oppressed and their leaders educate the leaders themselves to equality they stand only to alter the style of popular servitude. Therefore, the pragmatics of communicative competence belong to a larger critical theory of socialist education which is broader than either the cognitive theory of proletarian consciousness or a linguistic approach to the critique of ideology. Socialist education cannot succeed except where men *love* the world and the society of men. The pedagogy of the oppressed therefore cannot be one-sided or manipulative nor can it be fostered by men who lack *humility* and are arrogant and domineering. Socialist humility in turn has no place where a few men set themselves up and are unwilling to bend in the service of the people because they lack *faith* in the partnership of the oppressed. However, critical socialist pedagogy is more than a prescriptive ethic. As outlined by Paulo Freire, whose work I regard as a practical extension of Habermas' ideal of emancipatory communicative competence, the pedagogy of the oppressed involves an applied method of entry and organization of local communities whose critical tasks are opened up through the *generative linguistic themes*—slum, development, water, infant mortality—of daily life in the colonial world.[13] It would be an important and relatively easy intellectual task—as easy as in practice it is dangerous—to extend Freire's methods of linguistic "conscientization" to the education of the internal proletariat of industrial and urbanized societies where the challenge does not derive from absolute illiteracy but from the relative illiteracy that results from the increasing technical and professional practice of social change and reform. In particular, the task of a critical theory of communicative competence under these conditions would require the development of adequate ethnographies and dramaturgies of lay and professional "accounting" procedures as competing epistemic and organizational strategies of communal order, conflict, and change. It is along these lines that I believe critical theorists must work if those systematically excluded from having a voice in the legitimation processes of our society are to realize their potential for citizenship.

What this task involves is, as Marike Finlay has developed with remarkable force, a discursive shift in communication studies away from the content of new technologies, data banks, hardware and networks, to question how these procedures operate to produce closure/exclusion of various users, programs and information. The contrast between this discursive approach and traditional studies of communications technology in abstraction from the framework of a critical theory of history and political economy is nicely formulated in Table 5.1 from her work[14] on "powermatics."

Critical theory cannot remain an academic preserve. It must be broad enough to share the historical burdens of labor anywhere in the world. It therefore cannot weaken its grasp of political economy. At the same time, however, critical theorists must commit themselves to the ordinary work of a social science that will deliver competent ethnographies of the practical exigencies of daily living in the modern world.[15] The accumulation of such knowledge requires that critical theorists develop a broad pedagogy that would restore the political relation between science and commonsense knowledge and values. At this level a critical theory of communications must begin work wherever the opportunity affords itself, in quite local circumstances and over simple matters in schools, hospitals, supermarkets, the arts and social services as much as in the factories. Only after a long period of such work ought we to climb once more the ascending heights of theory.

Table 5.1.

Traditional approach	Alternative approach
Disintegrated: lack of integration of techniques, functions and issues;	*Integrated*: view based on common denominator of discourse, i.e., a satellite and a smart toaster are both practices of social discourse;
Empiricist's paradox: insistence upon the criterion of empirical observation and its adequacy, as a representation of the world—objectivity of science;	*Epistemology based on discourse*: the recognition of the perturbation of empirical results; and the belief that "facts" and "events" are discursively manifest and co-constituted;
A-historical: futurological bent and evolutionary view of history which ignores the past and refuses to answer to the specific demands of the present context of communication practice;	*Historical contextualization*: attempt to fulfill the need for a historical study which contextualizes new communications technology in relation to the past ways of conceptualizing it as well as in relation to the demands made by current social conditions—i.e., a form of discursive materialism;
Idealization/abstraction: idealization of technology to an abstraction from the concrete context which is formative of both the uses and the very structure of new communications technology;	*Contextualization*: attempt to develop a contextual theory of technology and society whereby technology is neither "inherently neutral" nor "inherently value-laden" but where it is studied as what grows to be, to be used, and to be judged within specific contexts;
A priori assumption of context: an a priori view of what the context of new communications technology is once and for all, be it one of social responsibility or exploitation;	*Changing context*: constant reconstitution of contexts is based on a study of the traces of context left by past and present discourses;
Out of focus: no theoretical framework by which to judge the validity and consistency of statements and positions re host of new communications technology;	*In focus*: bringing to light the dominant focus of power and social control; eventual search for an alternative focus: democracy and participation;
Claim to objectivity: the positivist belief that the theory of	*Affirmation of interests*: all theory, including positivism, is value-laden or

communication is neutral and objective, descriptive as opposed to prescriptive;

linked to ideology and therefore these values should not be occulted but rather exposed and defended as discursive and knowledge guide-lines for action, i.e., for technological intervention and the legitimation of alternative configurations of power;

Reaffirmation of status quo: the acceptance of the status quo because it is 'fact' and because the *possible* is not a viable object for empirical knowledge;

Potential: the criticism of existing practices of communication in view of postulating the possible practices in a possible state of affairs;

Closure within an accepted discursive space: no attempt to reflect upon the logical space within which their positions are enunciated; no misery due to extreme cliche of existing texts; no attempts to overcome the limits of their own critical discourse.

Relativization of logical space: a constant attempt to dejargonize the cliches surrounding the debate, to expose the mythification as well as to find discourse with which to talk about constraints or procedures of one's own discourse while seeking to transform those which deny potential alternatives.

NOTES

1. In constructing these propositions regarding the communicative relationship between political legitimacy, goal articulation, and resource allocation, I am drawing upon R.K. Merton, "Social Structure and Anomie," in his *Social Theory and Social Structure* (New York: The Free Press of Glencoe, 1963), pp. 131-160, 161-194.
2. Claus Mueller, "Notes on the Repression of Communicative Behavior," in Recent Sociology, No. 2, *Patterns of Communicative Behavior*, Hans Peter Dreitzel, ed. (New York: The Macmillan Company, 1970).
3. John O'Neill, "Eye Work in the City: The Natural Production of Civic Order," published in French as "Lecture visuelle de l'espace urbain," in *Colloque D'esthetique appliquée á la création du paysage urbain* (Paris: Copedith, 1975), pp. 235-247. I have developed more fully the study of bodily orders in political economy in my *Five Bodies: The Human Shape of Modern Society* (Ithaca: Cornell University Press, 1985).
4. Max Weber, "The Types of Legitimate Domination," in *Economy and Society*, Vol. 1, G. Roth and C. Wittich, eds. (New York: Bedminster Press, 1968), p. 212.

5.   Jürgen Habermas, *Legitimation Crisis*. Trans. by Thomas McMarthy (Boston: Beacon Press, 1975); Claus Mueller, *The Politics of Communication: A Study in the Political Sociology of Language, Socialization and Legitimation* (New York: Oxford University Press, 1973).

6.   See Chapter Two.

7.   *Legitimation Crisis*, pp. 36-37.

8.   Carl Kaysen, "The Corporation: How Much Power? What Scope?" and Norton Long, "The Corporation, Its Satellites and the Local Community," in *The Corporation in Modern Society*, Edward S. Mason, ed. (Cambridge: Harvard University Press, 1961), pp. 85-105, 202-217.

9.   See the argument between Habermas and Luhmann in *Legitimation Crisis*, Pt. III, Ch. 5, "Complexity and Democracy."

10.   O'Neill, "Authority, Knowledge and the Body Politic," in *Sociology as a Skin Trade*, pp. 68-80.

11.   Jürgen Habermas, *The Theory of Communicative Action*, trans. Thomas McCarthy (Boston: Beacon Press, 1984, 1987), 2 vols.

12.   Jürgen Habermas, "Toward a Theory of Communicative Competence," in *Recent Sociology*, No. 2, p. 144.

13.   John O'Neill, "Decolonization and the Ideal Speech Community: Some Issues in the Theory and Practice of Communicative Competence," in *Critical Theory and Public Life*, John Forester, ed. (Cambridge: The M.I.T. Press, 1985), pp. 57-76; Lee Thayer, *On Communication: Essays in Understanding* (Norwood: Ablex, 1987), "Part III: Some Ethical/Moral Issues," pp. 127-158.

14.   Marike Finlay, *Powermatics: A Discursive Critique of New Technology* (London: Routledge and Kegan Paul, 1987), pp. 336-337.

15.   John O'Neill, "Critique and Remembrance," in *On Critical Theory*, John O'Neill, ed. (New York: Seabury Press, 1976).

PART II

LIFE-WORLDS AND THE
THE MEDIATION OF BIO-POWER

CHAPTER 6

Tele-vision and the Nationalist Platitude: Reflections on the Canada/USA Watch

No doubt we are sensible in giving names to places: Canada; the Argentine. But we would also be sensible to remember that the land we have given these names to, and all but the relatively very small human population, wears these names lightly.

James Agee

Although the United States is ungenerous in leaving Canada off its map, Canada and the U.S. are generally to be seen in terms of one another. We enjoy a certain marriage—whose vows have recently been renewed under the so-called Free Trade Deal. Since both Canada and the United States are too large to be visited, one is obliged to see them from afar. For this purpose "tele-vision" is essential even though it makes it difficult to disentangle what is specific to each culture. If Canadians insist upon a secure access to the American market, it is because they are afraid to lose the mirror in which they see themselves. Nothing depresses Canadians so much as a barrier between themselves and the United States. The alternative is to try to love Mother England or the narcissistic French. To the new Canadian immigrants neither alternative is as attractive a mirror since America is the mirror in which they have been accustomed to view the world.

111

Although Canadians are reluctant to ascribe to themselves the metaphysical luxury of a soul, they nevertheless defend something called the "Canadian mind" whose nature may be guessed from the present essay and the following one on empire and communications (Chapter 7). The Canadian mind is not possessed of any grandiose charter in which it celebrates the natural right of peoples everywhere on earth to abandon the culture they have enjoyed for thousands of years in order to become Canadian. Even our new Charter of Rights operates more like a franchising system for front store lawyers than a beacon to the world's immigrants. The invidious tone of these remarks should be recognizable as "Canadian," that is, inspired by considerable comparison, for better or worse, with the United States. They are, then, part of the marriage between us; they belong to the daily "sitcom" of Canada-U.S. relations. Our situation, however, can be expressed rather more urgently. No one in Canada is better able to capture this urgency than my colleague Arthur Kroker whose work must be considered on its own terms by anyone interested in the issues that will be sketched here. Kroker employs the metaphor of marginality to capture the pathos of the Canadian mind whereas I employ the imagery of an uneasy marriage. The difference between us grows less when one considers the issues we take to be at stake. Kroker puts things as follows:

> the Canadian mind is that of the *in-between*: a restless oscillation between the pragmatic will to live at all costs of the Americans and a searing lament for that which has been suppressed by the modern, technological order. The essence of the Canadian intellectual condition is this: it is our fate by virtue of historical circumstance and geographical accident to be forever marginal to the "present-mindedness" of American culture . . . and to be incapable of being more than ambivalent on the cultural legacy of our European past.[1]

In this chapter I attempt to bring together a display as well as an analysis of the elements of fact and myth which play their part in Canadian nationalism. My method is phenomenological inasmuch as it does not attempt to reduce the constellations of nationalist myth and consciousness to either ideology or illusion. At the same time, my analysis of the basic features of the Canadian political economy is Marxist while allowing as much weight as possible to the cultural and historical overdetermination of Canadian relations with the United States. My efforts to combine the methods of Marxism and phenomenology should also be taken as a commitment to a particular style within the possibilities of Canadian political economy. It is natural to us to draw upon the tradition of European social science, to be obsessed with the power and affluence of its American profession, and to hope for some way of our own cultural development. I have deliberately set aside any argument from social development theory that nationalism is a regressive movement that can only delay the necessary trend towards world modernization. Anyone concerned with this view can find it in its most flamboyant form in the works of Marshall McLuhan, with whom Canadian intellectuals have for a time flirted but eventually abandoned in favor of allegedly more stable marriages with professional disciplines. In the end I believe my own position abides by the Canadian experience through its openness to the variety of arguments which bear upon the nature of Canadian nationalism and the occasion it offers for the orientation of social thought.

I

Consider what is sometimes called the *Canadian fact*, namely, the overwhelming American dominance of the Canadian economy and its implications for Canadian sovereignty and cultural identity. It soon becomes clear that the statistical arguments over Canadian foreign dependence are not easily evaluated without making further assumptions about the degree of interdependence between national sovereignty, economic independence, foreign policy, and national identity. All the same, the sailing of the Esso tanker *Manhattan* through the Northwest Passage hardly required much guesswork at that time over the nature of American determination to exploit world resources, careless of the political and ecological consequences of America's imperial needs. Of course, imperialist practice in the end may well be redefined in terms of more or less mutual arrangements between the United States and Canada, as in what we now call the Free Trade "Deal." But for this to be likely it is necessary to understand the belligerent stance of the United States with regard to what it calls "world resources," a usage which is intended to assure America's own access to these resources wherever they may originate. Any argument which speaks about continentalism and economic rationality in abstraction from the world strategies of the multinational corporation and American military postures must be regarded as American pie. The Shultz Report made such considerations clear in its subtitle: "A Report on the Relationship of Oil Imports to the National Security." The Report evaluates the security of the United States in shamelessly nationalist terms even in the midst of futuristic scenarios of world destruction.

On the other hand, it can hardly be claimed that Canada's experience of American economic domination has driven us into Third World revolution, although Canada shows occasional interest in the Mexican response to its problems with the American border. Everywhere one turns, the facts of American exploitation are clear enough to raise the voice of national indignation. It is amusing to find American observers alarmed by the "doctrinal nationalism" of the Canadian media. The passage of the *Manhattan*, and the aggressive overtures of American Secretary of the Interior at that time Walter Hickel did, however, provoke J.J. Greene, then Canada's Minister of Energy, Mines and Resources, to reject the American way of life, at least in Denver. Indeed, all political parties have felt the necessity of hawkish responses to the American invasion of the Canadian economy, in particular, the Waffle group in the New Democratic Party when led by Watkins and Laxer, as well as the non-party Committee for an Independent Canada. But these pronouncements are usually softened if they result in an election victory.

The test of the Canadian response to American hegemony is ultimately the same test that the Canadian government faces in the invention of an integrated conception of socioeconomic, political, and ecological development which will give it better control over inflation, levels, and location of employment, foreign trade, and foreign policy Towards this goal there are a number of starting points, among which anti-American socialism is only one, although those who adopt it seem better able to tackle the illusions and myths which befog the trail leading to the peaks of political, economic and cultural sovereignty. Other competitors appear more modest. Meanwhile, their practicality, combined with sophisticated cultural despair, works inevitably in favor of continentalism at the expense of Canadian identity and unity. For, of course, the separatist strains in Canada are very much related to the unequal regional participation in the continental economy governed by the corporate agenda and liberal ideology. However, it is my own view that the branch plant consequences of the multi-national corporate technology and its investment practices, which have been so tirelessly documented,[3] are less important than the phenomenon of the dependence of the social and political structures of countries through which the American multinational corporation reproduces itself. This is not simply a matter of the obvious integration of multinational production and marketing processes, about which we have so much information, but also of the dependence and imitation of consumption patterns which are not suited to the needs of the dependent economy, rationalized as progress and modernity, combined with poverty and regional backwardness in the deep hinterland of the colonized economy. In other words, a major effect of the multinational corporation is to transplant its own class distortions of the national agenda upon its dependent hosts. Thus a *dependent class culture aggravates the metropolitan-hinterland contrast, forcing the definition of progress, modernity, and urbanity*

into a replication of the cultural dualism which is presupposed by the opera-
tions of the multinational corporation and hardly modifiable by liberal gov-
ernment allocations of social resources.

Surely one of the most important effects of the multinational corporation is the distortion of the dependent country's social energies through
paths of development which merely sediment the consumption style of its
ruling minority into the economic infrastructure, thereby replicating the production pattern of the relatively advanced economies. In this way, there is
guaranteed a *specious internationalism of the development process*, which
the local bourgeoisie appeal to as the path out of the backwoods into the
light of day. The structural reasons for the "irresistibility" of such modernization, which stem from the nature of the technological and investment
processes within the multinational corporation, are made to look like
inevitable forces of history which cannot be ducked and so must be
embraced, thereby legitimating the political economy of continental or global dependence. I shall return to this argument in my consideration of Grant's
lament for the Canadian nation.

Yet in Canada we have no revolution, the Canadian fact notwithstanding. Perhaps, then, we must look into ourselves instead of away from
ourselves and our preoccupation with statistical ventriloquy. It is hard to
establish the case for rape where the victim has such a long history of seduction and sellout. In Canada we are able to face the facts of exploitation by
the United States with liberal sophistication rather than with Marxist hyster-
ics. After all, Canada has no desire to be left on the shelf, an industrial virgin, excited by nothing but the unconsummated call of her nature lovers. For
conservation is a fate worse than penetration, even though the American
embrace may at times be a little rough. Moreover, the marriage is not without the consolation that in exchange for allowing the United States to win
our bread for us we may be able to civilize American husbandry into the
paths of a more stable, less violent way of life. This is a thought which is
generally a comfort to us and on occasion appears to be the essential wisdom of Canadian-American relations. It is the reward for the daily sacrifice
of marriage and dominance, although the wisdom and modesty of these
arrangements is often challenged by Canada's rebellious children, her artists
and intellectuals who cannot wait to be seduced and raped in the big cities
of New York, London, and Paris and then to record their poetic screams for
their small town countrymen, stifling in a backward nationalism.

In those calm days of the American Empire, before Vietnam and
before American children began to play politics in the streets, the intellectual roots of liberal sophistication rested on the myth of social science knowledge that we had escaped the old world history of oppressive national identities into the new world drama of personal identity. While the heart of the

American Empire was still Hollywood (it is increasingly transplanted into Toronto, Montreal, and Vancouver), we could all claim double citizenship in the dramas of the wild west, the Chicago gangs, and any one of the endless romances of the Hollywood stars. In many ways sociology, too, is an American medium and its message has not been different from that of Hollywood in the sense that it is as much a science of escape from poverty, ignorance, and oppression, as it is a science of social structure and individual relationships. In short, the older civilizations—and Canada, through France and England, could share this view—have regarded the American preoccupation with questions of personal identity as the essence of American innocence, unclouded by national politics and racial ambition. The price of a movie ticket appears to have been all that it cost for double citizenship. *Television confers upon us a permanent passport to American culture.*

But this is no longer an adequate picture. The tragedy of America is that its dream of personal identity has forced so many to become the misfits of a precariously homogenized culture of affluence that is entirely lacking any genuine private or public roots. Increasingly, American personal identity is submerged in the corporate face and its imperialist ventures. At the same time, social science knowledge has become self-conscious of its own Americanization and tries desperately to extricate itself through the boot strap operations of the sociology of sociology. This is not to overlook, of course, that the pragmatic liberalism of American social science has been deeply committed to the adventure of imperial identity in Asia no less than to American corporate delinquency at home.[4]

As James Laxer has observed:

In general American radicalism tends to be an inappropriate guide for Canadian radicals because it is conceived out of the conditions of the heart of the empire rather than the conditions of a dependent country. American radicals to a large degree are concerned with preventing or checking the exploitation of much of the world by their own country. This leads such radicals to be quite unconcerned about the effectiveness of American institutions. When Canadian radicals, influenced by the American New Left, adopt the same attitude to the effectiveness or survival of Canadian institutions, it has far different political implications. It benefits the world when American radicals challenge the right of U.S. institutions to continue their economic and military domination of people abroad. This has reinforced the New Left's distrust of organizations and institutions in general. A similar attitude to Canadian institutions, however, results merely in a further softening up of this country for American take-over.[5]

We need to exercise care in the argument over the relations between science, scholarship, and society. What students criticize as the lack of Canadian content, or the lack of relevance in their studies, is not especially a phenomenon of the Americanization of knowledge so much as the *professionalization* which also serves imperialism.[6] The professionalization of knowledge is the primary instrument of technique which is in turn the dominating force in the bureaucratic manipulation of persons as resources determined by status and office. The professional status of social scientists themselves has in turn depended upon their recognition as functionaries servicing the liberal theology of technique through which the industrial corporations reduce society and nature to an organizational strategy. There can be little doubt that, at least in Ontario, we have built monoliths to the Americanization of knowledge stampeded by the moon game. I do not think we shall have understood this properly until Canadian universities have faced up to the national need not only to accommodate open learning but also to institutionalize advanced knowledge devoted to rethinking our daily practice.

II

I want now to pursue the question of Canadian nationalism by turning from the consideration of the intellectual background predisposing to the myth of the nation to the consideration of the problems of national myth-making itself. The creation of the myth of Canadian national identity is perhaps the most painful and potentially divisive experience to be faced in the university community, if not the nation itself. For all that we know of the biographical and cultural influences upon knowledge, we are still inclined to believe that these are effects which we should strive to minimize rather than to encourage positively. A nationalism of knowledge presents itself as a horror to the minds of university men and women who have worked in the long tradition of western civilization, happy indeed to make for some betterment of life but unable to understand what any national contribution might mean that did not offer itself to all men, or that could not withstand comparison with what is done elsewhere. There can, however, be very little doubt that nationalism is a powerful political sentiment which many in Canada, as elsewhere in the world, feel would provide a viable medium for the solution of our constitutional issues of Confederation, the union of French and English Canada, as well as offering a vehicle for an offensive economic strategy against continentalism or American imperialism.

We need to be as honest as we can with our emotions, for they are divinities that shape our ends, rough hew them how we will, if you will pardon my Shakespeare. My colleague Lionel Rubinoff has argued, for example, that a sense of national purpose finds no authentic response in him as a Canadian.[7] He prays that he might be preserved from such a fate. His thesis is that a sense of manifest destiny only comes with the anxiety of power and in turn feeds the power that breeds it. Like many other Canadian intellectuals, he wants no Canadian "Way" of life, no "Dream" to feed the engines of technology and power. Yet I know he, like most of us, can feel pride in Canada, whether it is in an occasional sports trophy, not to mention literally beating the Russians at hockey, or in something as subtly moral as the effects of a harsh climate on building character. For one can become a nationalist even in a canoe or on skis. I am sure that Lionel and I often yielded to chauvinistic comparisons between Canada and the United States, or to the moralist hope that our children would grow more straight in the cold of Ontario than in the heat of California—and, for myself, I think the sole reason why I watch television is to feed such chauvinistic comparisons, sentimentalizing Canada in the stupid hope that the United States will simply sink under the weight of its terrible problems and leave us all be. All the same, I think I can live with these mixed feelings about nationalism; for I cannot see how a man or woman can live without some kind of love for the place in which he or she lives and works and feeds a family. I can understand a feeling of solidarity with others who share this life and of loyalty to those who in the past built its ways. I do not believe that such nationalist sentiments are necessarily tied to militarism and the demand for manifest destiny. Moreover, I believe such feelings are not to be reduced, despite what liberals make of them, to the bad antitheses of Canadian inferiority feelings with regard to Britain, or to an empty internationalism that merely covets American strength.

To the prophets of electric technology, nationalist sentiments must surely appear to be a backward medium when in reality nationalism may be the perfect tribal response to the need to preserve local color, to humanize international robotry, and enable us to find ourselves in the electric grid. The Canadian inventions of Marshall McLuhan serve us poorly if they are used to recommend a total externalization of mind through an international network owned by Americans, instead of serving the argument for national planning as a path to cool individual life styles. The latter course is surely one that Abe Rotstein[8] and Walter Gordon[9] have labored to make less precarious. Yet the myth of technology, despite what it owes to Canadian electrification and McLuhanite promotion has still in Canada to fight the myth of the land. "But while each of these myths serves respectively the cause of continentalism and nationalism, the land is divided against itself and may foster either cause."[10] *The geography of Canada supports two conflicting identities*. To some, Canada exists because it is a distinctive northern land, whose harshness is the strength of Canadian purpose and the basis of its East-West saga, of the fur trade, railways and the new electric networks of radio and television. To others Canada is a geographical illusion whose boundaries, at any rate to the South, are more honored in the breach.[11] Canada's ambiguous geography underlies the competition of the nationalist and continentalist versions of Canadian history, to which some reference must be made before we turn to these alternatives as they play their part in current Canadian political controversy. It needs to be said that the practice of Canadian sociology in general and more particularly anyone espousing Marxist analysis cannot afford to overlook the historical, regional, and cultural factors shaping Canadian facts. To the extent that we are mesmerized by American sociology, we become short on historical meaning and immune to specific cultural practices neglected in favor of some potential scientific generalization.

Every people faces the problem of survival and the story of its successes and failures is woven into the history and politics of the land which is the setting for the nation's deeds. The land is an essential element in the myth of survival, of hard times, of national growth and development. The Canadian land is an especially vibrant theme in the historical and political consciousness of its politicians and historians. The issues of nationalism and continentalism spring naturally from the great East-West expanse of Canada and from its bottomless southern boundary with the United States. It has been the great contribution of Harold Adams Innis to set the political and constitutional development of Canada in the framework of her economic geography. Innis argued that the basic fact of Canadian economic history has been its ties with the European economy based upon the East-West axis and the development of the fur trade, fish industry, the Canadian Pacific Railway and the St. Lawrence waterway. In this way he provided a corrective to previous historical writing based upon the themes of Canada's independence from Britain and her inevitable integration with a North American continental economy:

Throughout the economic history of Canada, the dominance of water transportation in the Maritime Provinces and the St. Lawrence has accentuated dependence on Europe for manufactured products and for markets of staple raw materials. The fur trade was followed by the timber trade and agricultural products. Concentration on staple commodities was accentuated by the migration of technique from the United States. As the export trade in staples from the United States to Great Britain declined in importance, the Canadian trade in staples was encouraged. The fur trade was strengthened by American aggressiveness and technique, the timber trade shifted from New England to New

Brunswick and the St. Lawrence, the fishing industry migrated from New England to Nova Scotia, agriculture, in the production of wheat in Ontario and the Prairie Provinces and in dairying, benefited from the contributions of the United States. The dependence of Canada on Great Britain was accentuated by the United States indirectly and by British and Canadian policy directly. European markets and European capital dominated Canadian economic development through the background of water transportation.[12]

However, as Innis noted, there occurred a shift in the axis of the Canadian economy, away from a transcontinental, confederational pattern into a continentalist and regional organization that is potentially destructive of Canada as a confederate nation.

The end of the period of expansion based on the St. Lawrence and trade with Great Britain coincided roughly with the achievement of dominion status which followed the Great War and which was marked by the Statute of Westminster. The end of the struggle for control over external policy has been accompanied by the increasing importance of regionalism evident in the growth of the powers of the provinces. The cultural features in terms of language, religion, metropolitan and political organizations based on the peculiarities of staple trades from various regions of Canada to Europe, which provided the basis of the provinces in the British North America Act, have hardened and been strengthened by the decline in the influence of the St. Lawrence as a centralizing factor in the Canadian system. The expansion of provincial powers, conspicuous in New Brunswick, Ontario and Quebec, has been scarcely less evident in Manitoba, Alberta and British Columbia. The decline in commercialism which accompanied the rise of free trade advocated by Adam Smith and his disciples left a structure which moulded the growth of capitalism (sponsored by those who paid lip service to Adam Smith) and hastened the growth of protectionism. The extension of the American empire, the decline of its natural resources, and the emergence of metropolitan areas, supported capitalist expansion in Canada and reinforced the trend of regionalism. The pull to the north and south has tended to become stronger in contrast with the pull east and west. The British North America Act and later decisions of the Privy Council have strengthened the control of the provinces over natural resources such as minerals, hydro-electric power, and pulpwood on Crown levels, resources which have provided the basis for trade with the United States and for investment of American capital.[13]

Thus Canadian nationalism is rooted in the image of the land. In turn, the land is essential to the imagery of Canadian survival and we must now turn to our great national debate on this theme. It has been argued that the Canadian bourgeoisie deliberately sacrificed the revolutionary features of old style and new style colonial nationalism, which would have demanded genuine national building, to its own brand of nationalist rule, preserved by keeping French and English Canada apart, except for purely opportunist appeals to national unity in times of liberal crisis.

> The charge that Canada has become *de facto* a region of the United States is not new, and on one level it does not take Canadian liberals by surprise. For the bourgeoisie who have guided Canada into a deepened colonialism the wide spectrum of dependency is not a disgrace, nor is it regarded unfavorably. Near-prosperity, the illusion of prosperity, the spill-over from imperial prosperity have been reason enough to legitimize their policies in the political and economic spheres. Their success in living off the British and American empires has given the bourgeoisie enough political power to withstand the opposition from a long line of radical and radicalizing movements for social change. In the past, the politics sustained by regional disparity, class struggle, farmer militancy and Quebec separatism never reached ascendancy. Hence, for good reason, the bourgeoisie were able to pursue policies which guaranteed them their place as managers and middlemen of a colonial economy.[14]

To the outside world our great national debate will be associated with the boy who kissed the girls, if not with the more serious national discussion to which Trudeaumania was a romantic prologue. In comparison with what went before, the promise of the Trudeau adventure seemed well worth the plunge into a youthful identity crisis, constitutional debate, the flag, bilingualism, Quebec separatism, left nationalism, and anti-Americanism. Events have been hard on this romance and its entanglements. It is not just that we discovered the Prime Minister's real age, nor even that there was a married man behind all that gaiety. There was the utter failure of Trudeau's Just Society to show the weakness of liberalism, the War Measures Act of October 1970 and a growing awareness that Canadian society, whatever its justice, is increasingly less our own and more and more Americanized. Trudeau's own political response to these issues was to fight Quebec nationalism as well as American imperialism with his concept of Canadian federalism. His ideal was a nation-state which could contain the two nationalisms of French and English Canada, along with a mosaic of other national immigrant cultures, without leading to either a homogenized or nationalistic society. Trudeau's critique of the majority and minority nationalisms that threatened to divide Canada was based upon his special conception of political consensus. The latter is often misunderstood, since Trudeau's manner appeared to waver between authoritarianism and nonchalance when faced with separatist declarations of nationalist purpose. Trudeau's theory of nationhood, however, was based upon a conception of everyday political consensus which may well have its roots in a historically nationalist determination but cannot be sustained by such impulses under the complex conditions that confront modern sovereign states. A modern state, Trudeau argued,[15] will do everything in its power to cultivate national sentiment, culture, ritual, and welfare—but it cannot be ruled by nationalism. For the modern state is best ruled by reason, although it must know how to serve the emotional needs of its people.

Federalism, by its very nature as a compromise, represents the best example of political reason which seeks to sustain consensus. Federalism, unlike nationalism, restricts the area over which consensus is imperative and in general relies upon the relative attractiveness to minority groups of membership rather than separation. Thus the modern state must be ready to invest a great deal of energy and resources in nationalism *at the federal level* in order to reduce the appeals of separatism. Trudeau's conception of federalism, then, was perfectly cognizant of the right to national self-determination—indeed, it is built upon the achievement of the sense of such determination, which is an emotional good, realized, however, within the rational and pragmatic framework of federalist politics. Moreover, it is part of this conception that it should prevail in international politics over the Christian conception of sovereignty, which could only be dominant in a world "crippled by the ideology of the nation-states and sustained by the heady stimulant of nationalism." Trudeau rejected all forms of nationalism, whether Quebecois separatism, Anglo-Canadian nationalism, or any combination or variant of these determined by anti Americanism.[16] His reasons were ultimately the same as those of Frantz Fanon, namely, that the nationalists are ultimately petit bourgeois counterrevolutionaries. Theirs is a wigwam complex, satisfied to demand control over the political means of production. Nothing proved this more than the Parizeau/Landry's push for separation, trashing new immigrants and native peoples.

III

The focus of my attention for the remainder of this chapter must be upon George Grant's reflections upon Canadian survival and the North American empire. The necessity here is my fascination with the depth and eloquence of an argument at the very heart of Canadian survival. It is also striking that Grant was a philosopher and not a sociologist—once again raising the question of our borrowed definitions of sociological practice. Grant's lament for Canadian nationalism is remarkable in the way that it confuses liberalism and Marxism as vehicles of the technological and world domination directed by the multinational corporation. It is interesting to observe, and by that I do not mean damaging, that Grant came from the earliest wave of American draft dodgers—the Loyalists who left America because its Revolution seemed to them to destroy the moral unity of the English speaking, if not Christian, world. However, what the Loyalists found was that Canadian liberalism leaves the door open to a technological continentalism that has escalated the war of values directed from Hollywood, Disneyland, and Madison Avenue beyond anything plain revolutionaries might have achieved.[17] The early Loyalists placed great hopes in the British Commonwealth as a moral bulwark against American Babbitry, individualism and progressivism. But by 1945 this hope was quite forlorn. The Americanization of Europe became increasingly a fact and Britain found herself hardly any stronger than Canada to face the challenge of American economic supremacy.

Grant's *Lament for a Nation* is in reality an essay on the withering away of history even though some might consider it, together with his *Technology and Empire*, as a Canadian contribution to the renewal of the philosophy of history. Certainly, Grant's thought is historical and moral in the grand conservative tradition. Unfortunately, his experience is modern. Indeed, in Canada Grant stood on "the sharp edge of modernism" and with it the temptations of nostalgia and alienation. Yet Grant argued the impossibility of conservatism in the modern world, a world which he understands, however, to be undistinguishably the world that the liberals and Marxists have made. "The impossibility of conservatism in our era is the impossibility of Canada," says Grant. "As Canadians we attempted a ridiculous task in trying to build a conservative nation in the age of progress, on a continent we share with the most dynamic nation on earth. The current of modern history was against us."[18]

To the extent that Canadian independence was based upon the west-east flow of trade, it was tied to the loyalty of the national bourgeoisie in whom the decline of patriotism was an inevitability of the continental shift of trade which we have described earlier. All the same, Canadian nationalism was more than a front for bourgeois interests, although we have seen that questioned by Daniel Drache, and Grant himself saw that political conservatism plays into the hands of capitalism.

To put the problem directly in terms of our contemporary society: there can be no doubt that we all have need of a proper conservatism, an order which gives form to persons, to families, to education, to worship, to politics and to the economic system. Yet to express conservatism in Canada means *de facto* to justify the continuing rule of the businessman and the right of the greedy to turn all activities into sources of personal gain. The conservative idea of law has often been in the mouths of the capitalists, but seldom in their actions. Their economic policy has been the denial of order and form. Thus it is almost impossible to express the truth and conservatism of our society without seeming to justify our present capitalism.[19]

Grant nevertheless argued that the essence of Canada lies in the value that it places upon a society with a sense of public order and restraint, with a reliance upon tradition and a willingness to use government to control economic life in the interest of the public good. The foundations of this conception of political life are ultimately religious and never ideological. "To lament is to cry out at the death or at the dying of something loved. This lament mourns the end of Canada as a sovereign state. Political laments are not usual in the age of progress, because most people think that society always moves forward to better things. Lamentation is not an indulgence in despair or cynicism. In a lament for a child's death, there is not only pain and regret, but also celebration of passed good.[20] Because of this belief, Grant searched French Canadian Catholicism as much as his own Loyalist Anglicanism for a bulwark against the monolithic ideology of progress and freedom in which liberalism and Marxism are merely bubbles in a wave. Yet he was rightly impatient with Quebec's attempt to combine a semi-socialist state with provincial autonomy in order to preserve its culture.

Provincial control of economic development is not only useful for French Canadian nationalism but also for international capitalism. Any federal system of government strengthens the power of the corporations. The division of power weakens the ability of public authority to control private governments; the size of the provinces allows them to be controlled by private economic power. The espousing by American or Canadian "conservatives" of greater authority for the local states has always a phoney ring about it, unless it is coupled with an appeal for the breakup of continental corporations. Decentralized government and continental corporations can lead in only one direction.[21]

The passing of the Loyalist and Catholic traditions of conservatism upon which a French and English Canadianism might have flourished is the result of the dominance of the great platitude, as Grant calls it, of modern progress which makes all local cultures an anachronism. The argument that Canada, a local culture, must disappear can, therefore, be stated in three steps. First, men everywhere move ineluctably toward membership in the universal and homogeneous state. Second, Canadians live next to a society that is the heart of modernity. Third, nearly all Canadians think that modernity is good, so nothing essential distinguishes Canadians from Americans. Together, these three propositions contain everything that is soporific and lethal in the Canadian fact. They are easily swallowed. Perhaps to some they generate a certain preservation, at least in their contempt for the daily practice which they observe in others, if not themselves. And yet it is clear that these apparent facts, so far from containing a simple description, are largely a prescription or a diagnosis that can be challenged. Grant, however, preferred to drink the cup to its bitter dregs.

I cannot develop to any length the objections to Grant's three theses. But, to begin, I might observe that not every experiment in federalism and government control, given the cultural background of Canadian leftists and conservatives alike, can be identified with the adulation of the universal and homogeneous state, better described by Marcuse as one-dimensional society. As for Canada's need to recognize that it lives next door to the United States, nothing follows as to her unambiguous recognition of American modernity—the slums of homicidal Detroit and the filth of Buffalo are hardly welcoming entrances to the modern world on our borders. Again, Grant leaves the notion of modernity unexamined, except to identify it with progress at the very time when the costs of modernization and progress make many people feel that these are not directions, but rather the great hollow and vacuum in our lives, our communities and cities. Thus, with regard to Grant's third point, there are many Canadians who do question modernity, though he may be right that they are not to be found among those in power. But if he was speaking of the apathy of the masses Grant merely confused the price they have to pay in order to have anything at all, not to see their children die or to be broken with endless toil.

But Grant knew this. He did not lightly question the modern age. He understood that for the mass of men life, even with its present dangers, would be incomparably more brutal and short were it not for modern technology and the state. And yet he argued that it may be that we shall have to question the basic political philosophy that underlies the modern age and its identification of the good with individual freedom based upon the technological domination of nature and the minimization of social and political control. Thus in *Technology and Empire* Grant's argument deepened. What previously was lament became a great North American celebration of the primal impulses of "English-speaking" Protestantism connecting with the empiricism and utilitarianism of Baconian science, shaping the souls of men to this land of ours in a pattern of mutual domination and practicality:

> Now when from that primal has come forth what is present before us: when the victory over the land leaves most of us in metropoli where widely spread consumption vies with confusion and squalor; when the emancipation of greed turns out from its victories on this continent to feed imperially on the resources of the world; when those resources cushion an immense majority who think they are free in pluralism, but in fact live in a monistic vulgarity in which nobility and wisdom have been exchanged for a pale belief in progress, alternating with boredom and weariness of spirit; when the disciplined among us drive to an unlimited technological future, in which technical reason has become so universal that is has closed down on openness and awe, questioning and listening; when Protestant subjectivity remains authentic only where it is least appropriate, in the moodiness of our art and sexuality, and where public religion has become an unimportant litany of objectified self-righteousness necessary for the more anal of our managers; one must remember now the hope, the stringency and nobility of that primal encounter. The land was almost indomitable. The intense seasons of the continental heartland needed a people who whatever else were not flaccid. And these people not only forced commodities from the land, but built public and private institutions of freedom and flexibility and endurance. Even when we fear General Motors or ridicule our immersion in the means of mobility, we must not forget that the gasoline engine was a need-filled fate for those who had to live in such winters and across such distances. The Marxists who have described the conquest of the continent as an example of capitalist rape miss the substance of those events, as an incarnation of hope and equality which the settlers had not found in Europe. Whatever the vulgarity of mass industrialism, however empty our talk of democracy, it must not be forgotten that in that primal there was the expectation of a new independence in which each would be free for self-legislation, and for communal legislation. Despite the exclusion of the African, despite the struggles of the later immigrant groups, the faith and institutions of that primal encounter were great enough to bring into themselves countless alien traditions and make these loyal to that spirit. To know that parents had to force the instincts of their children to the service of pioneering control; to have seen the pained and unrelenting faces of the women; to know, even in one's flesh and dreams, the results of generations of the mechanizing of the body; to see all around one the excesses and follies now necessary to people who can win back the body only through sexuality, must not be to forget what was necessary and what was heroic in that conquest.[23]

Grant's testament to the affinity of the Protestant ethic with the spirit of capitalism remains one of the most eloquent formulations of the Weberian hypothesis anywhere. Canadian sociologists might well ponder it rather than the American version they usually rely upon. But it surely made it difficult for Grant to have his lamentations on the separation of necessity and goodness once he had joined with Marx and Weber in the exposition of the moral basis of modern technology and its expression in a system of rights, exchange and contract. The weakness in Grant's analysis is that he did not pursue it in terms of its Hegelian and Marxist sources.[24] Grant identified technology and freedom and attributed this liberal identification to Marx. But the Marxist identification is between reason and freedom.

Marx continues the critique of bourgeois civil society begun by Hegel, in *The Philosophy of Right*, to show that the liberal notion of subjective freedom in the context of unequal economic power is the key mechanism of the subordination of rational politics to the forces of capitalist society. At the stage of monopoly capitalism, or late modernity, as Grant might say, the rhetoric of individual freedom and equality no longer swells into a coherent political ideology as it once did in the antifeudal strategy of bourgeois and proletarian emancipation. The liberal theory of society and the individual was plausible so long as the historical situation which liberalism presupposed effectively linked its vocabulary of motives with typical contexts of action. The liberal image of man, its possessive individualism,[25] is the result of the imputation of the properties of market society to the psychic life of the individual. Once the evolution of market society is determined by the agenda of corporate capitalism, the vocabulary of liberalism, as well as Grant's loyalism, merely evokes lost contexts, arousing a nostalgia haunted by the loss of meaning.

Grant failed to understand what he called the tyranny of the modern universal and homogeneous state because he uncritically tied it to the process of technological domination. For the same reason, he was condemned to lament the destruction of Canada as a political and cultural entity. The processes of homogenization and continentalization are, as I have argued above, basic to the organization of the socioeconomic space of capitalism and the poverty of the political imagination of our times. The traditional antithesis of individual and state, society and state, public and private rights, conflict and order, no. longer serve to orient men's private lives toward their political contexts. Any political philosophy which does not question the nature of industrial technology and the power of large corporations to shape the national ecology and the psychic economy of individual lives is an empty word. The corporate economy stands between the individual and the state. Its power to determine the national life-style must be the focus of Canadian political economy and sociology working on behalf of

the same nationalization. However, the critique of the forces working to produce modern homogenization, or what Marcuse has called one-dimensional
man, must avoid both the *elitist fiction* that mass society is the cause of our
political troubles, a fiction with which Grant flirts, and the *liberal illusion*,
naively advanced by Galbraith, that pluralistic countervailing power is the
only viable form for political conduct.

We need also to understand how the imperatives of the corporate
agenda, both nationally and internationally, make "irrational" or violent any
attempts to alter its priorities in favor of the underprivileged, the laboring,
racial and linguistic "minorities" excluded from the process of affluent
homogenization. In Canada, as elsewhere, the enormous frustrations of
poverty are aggravated by its surrounding affluence, softened by a welfare
apparatus that has come under attack since global capitalism declared war
on public "deficits." The overprivatization of world resources worked by
monopoly capitalism results in the liberal stylization of the issues of power,
class and ultimately public indifference. The result is a North American funculture riddled with violence, racism, and colonial wars managed with very
low visibility, through the effects of regional and suburban segregation.
Where exploitation must be faced, it can be handled through the sentimen-
talities of charity, social reform and colonial aid which are the vicarious
counterparts of the affluent funculture and liberal politics.

The political culture of monopoly capitalism and its continentalist
or multinational style is able to marginalize all attempts to challenge its
massive dominance. It becomes very difficult for the workers of Quebec, the
Indians and militant labor in general to speak of justice and social outrage
that calls for violent overthrow in a society where such responses have been
made "irrational" and where political impulses have been reduced to the
passive expectations of the "goods" society. In such a context, violence is
made to appear "senseless," just as the exploitation which is its cause can,
for want of an adequate political analysis, be considered "unnecessary."
Such a context makes a *weak socialism* a Canadian fact but hardly a viable
national politics. Moreover, whenever it succeeds, it is always followed by
reprivatization. For the near future, we are likely to be divided against ourselves and our future between owners and nonowners of the national debt.
At this time, there is some anxiety that our concept of social services may
be thinned out as we are drawn into the American economy. To this extent,
class division will result, although we now lack any political party to exploit
such troubles.

NOTES

1. Arthur Kroker, *Technology and the Canadian Mind: Innis/McLuhan/Grant* (Montreal: New World Perspectives, 1984), pp. 7-8; Seymour Martin Lipset, *Continental Divide: The Values and Institutions of the United States and Canada* (London: Routledge, 1990).

2. *Ibid.*, p. 15.

3. Stephen Hymer, "The Efficiency (contradictions) of Multinational Corporations," *American Economic Review* (May 1970):441-448; L.W. Copithorne, "International Corporate Transfer Prices and Government Policy," *Canadian Journal of Economics* (August 1971):324-341; Kari Levitt, *Silent Surrender: The Multi-National Corporation in Canada* (Toronto: Macmillan, 1970).

4. Noam Chomsky, *American Power and the New Mandarins* (New York: Pantheon Books, 1969).

5. James Laxer, "The Americanization of the Canadian Student Movement," in *Close the 49th Parallel etc.: The Americanization of Canada*, Ian Lumsden, ed. (Toronto: University of Toronto Press, 1970), p. 278.

6. H.T. Wilson, "Continentalism and Canadian Higher Education," *The Canadian Review of American Studies* 1/2 (1970):89-99.

7. Lionel Rubinoff, "National Purpose and Ideology," in *Notes For a Native Land: A New Encounter with Canada*, Andy Wainwright, ed. (Toronto: Oberon Press, 1969), pp. 43-48.

8. Abraham Rotstein, *The Precarious Homestead: Essays on Economics, Technology and Nationalism* (Toronto: New Press, 1973).

9. Walter Gordon, *A Choice for Canada. Independence or Colonial Status* (Toronto: McClelland and Stewart, 1966).

10. Cole Harris, "The Myth of the Land in Canadian Nationalism," in *Nationalism in Canada*, Peter Russell, ed. (Toronto: University of Toronto Press, 1966), pp. 27-43.

11. Ramsay Cook, *The Maple Leaf Forever: Essays on Nationalism and Politics in Canada* (Toronto: Macmillan, 1971); John Kendall, "A Canadian Construction of Reality: Northern Images of the United States," *The American Review of Canadian Studies* iv (Spring 1974):20-36.

12. Harold A. Innis, "Significant Factors in Canadian Economic Development," in his *Essays in Canadian Economic History* (Toronto: University of Toronto Press, 1956), p. 208.

13. *Ibid.*, p. 209.

14. Daniel Drache, "The Canadian Bourgeoisie and Its National Consciousness," *Close the 49th Parallel*, p. 5.

15. Pierre Trudeau, "Federalism, Nationalism and Reason," in *Federalism and the French Canadians*, John T. Saywell, ed. (Toronto: Macmillan, 1968).

16. Trudeau, "Separatist Counter-revolutionaries," in *Federalism and the French Canadians*, pp. 204-212.

17. David V.J. Bell, "The Loyalist Tradition in Canada," *Journal of Canadian Studies* 2 (1970):22-33.

18. George Grant, *Lament for a Nation: The Defeat of Canadian Nationalism* (Toronto: McClelland and Stewart, 1965), p. 63.

19. Grant, *Philosophy in the Mass Age* (Toronto: McClelland and Stewart, 1966), pp. 107-108.

20. *Lament for a Nation*, p. 23.

21. *Philosophy in the Mass Age*, p. 77.

22. *Lament for a Nation*, p. 54.

23. Grant, *Technology and Empire: Perspectives on North America* (Toronto: House of Anansi, 1969), pp. 24-25.

24. Jean Hyppolite, *Studies on Marx and Hegel*, trans. by John O'Neill (New York: Harper & Row, 1973); *Hegel's Dialectic of Desire and Recognition: Texts and Commentary*, John O'Neill, ed. (Albany: State University of New York Press, 1996).

25. C.B. Macpherson, *The Political Theory of Possessive Individualism: Hobbes to Locke* (Oxford: Clarendon Press, 1962).

CHAPTER SEVEN

Bio-Communication: Empire and Biopower

I want to argue that the possibilities of modern biotechnology place us upon a frontier equivalent to that upon which Vico's first men found themselves.[1] *Today, we are called upon to rethink the human body.* But, as I see it, this involves more than an exercise in the new biology.[2] Rather, in rethinking the body we simultaneously rethink the *body politic.* This is because the new biology raises the threat of a *biocracy.* It thereby requires us to rethink our attachment to humanism and democracy. In order to show the urgency of the biopolitical issues on the new frontier of life, I think it is worthwhile to review the concepts of empire and communication in order to show how far the inscriptions of power recast man's sensory and cognitive experience. To do so, we may turn to Harold Innis inasmuch as he considered empire to be "an indication of the efficiency of communication."[3] That is to say, he thought empire and communication to be inextricable valorizations of power. Moving from Innis to McLuhan, we can see how it is that the inscription of power first creates a *sociotext,* so to speak, a network or tissue of power whose external manifestation is empire. At the same time, empire organizes the *sensus communis,* shifting the ratios of experience and sensibility, to rewrite the sociotext into *biotext.* We hope to clarify these notions

131

through a historical sketch, or a genealogy, whose usefulness lies only in its contribution to connecting the sciences of power and life.

BIOPOWER: THE BIAS OF COMMUNICATION

Innis and McLuhan inspire us to consider all political history to be inseparable from the history of biocommunication systems. Their work subverts the dualism in idealist and materialist historiography because they never consider human history as anything else than an *embodied history* inscribed upon the *communis sensus*. History is human history or *biotextual* because it alters our sensory and cognitive ratios but always in concert with the history of our land, its rivers and forests, its fish, fur and minerals.[4] It is the material history of these things that underwrites, so to speak, our mental and sensory histories told in our chronicles, monuments, and laws. None of this is caught in the reduction of communication to the techniques of information transfer. Thus in a later time, Innis and McLuhan re-echo Vico's claim in the *New Science* that men first thought the world with their bodies and only later did their sensory mind yield to the scriptural mind with which we have fashioned the conceits of rationalism:

> The human mind is naturally inclined by the senses to see itself extremely in the body, and only with great difficulty does it come to understand itself by means of reflection. This axiom gives us the universal principle of etymology in all languages: words are carried over from bodies and from the properties of bodies to signify the institutions of mind and spirit.[5]

In the light of Vico's axiom, it is necessary to argue that the ground of universal science is the world's body—upon which we inscribe our local logics and ontologies—and that *the world's body is the ecological setting of all our subrationalities*. We thereby ground the rational sciences in man's first poetic logic, in his poetic history and poetic economy. We do so, not to pit human reason against itself, but rather to fund the rational sciences in the memory of their first anthropogenesis.[6] Thus, as Durkheim and Mauss recall for us, all later logic is grounded in the act whereby the first men thought the order of things with their familied bodies, creating the world's first severe poem:

The first logical categories were social categories: the first classes were classes of men, into which things were integrated. It is because men were grouped, and thought of themselves in the form of groups, that in their ideas they grasped other things, and in the beginning the two modes of grouping were merged to the point of being indistinct. Moieties were the first genera, clans the first species. Things were thought to be integral parts of society, and it was their place in society which determined their place in nature.[7]

The first men thought society and nature with their bodies. Thus the first human world was a giant body whose divisions yielded the great divisions of the universe, of society and nature. These first imaginative universals generated an embodied logic of division and replication from which later rationalized modes of categorization could be developed. Thus the myths of the first men, so far from being poor science, are as Levi-Strauss[8] has also argued, the indispensable origin of human order and commonwealth without which rational humanism and scientism are impossible conceits:

It is noteworthy that in all languages the greater part of the expressions relating to inanimate things are formed by metaphor from the human body and its parts and from the human senses and passions. Thus, head for top or beginning; the brow and shoulders of a hill; the eyes of needles and of potatoes; mouth for any opening; the lip of a cup of pitcher. . . . All of which is a consequence of our axiom that man in his ignorance makes himself the rule of the universe, for in the examples cited he has made of himself an entire world. So that, a rational metaphysics teaches that man becomes all things by understanding them (*homo intelligendo fit omnia*), this imaginative metaphysics shows that man becomes all things by not understanding them (*homo non intelligendo fit onmia*); and perhaps the latter proposition is truer than the former, for when man understands he extends his mind and takes in the things, but when he does not understand he makes the things out of himself and becomes them by transforming himself into them.[9]

Let us recall Harold Innis rethinking the nature of empire and communication by rethinking its development in colonial Canada and from this margin deconstructing the monumental histories of the world's great empires. Just as he saw the fate of Canada pivoting upon its rival North/South, East/West axes to be a function of the changing role of the great staples of fish, fur, timber, and wheat, so he read the history of the great empires as similarly pivoted upon the staples of communication on papyrus, clay and stone, in print, books, newspapers, and radio. What is important in Innis' conception of the material history of power is that he never lost sight of the communicative struggle over monopolies of knowledge, or of the importance of regional resistance to communication empires that weaken democracy:

> Concentration on a medium of communication implies a bias in the cultural development of the civilization concerned either towards an emphasis on space and political organization or towards an emphasis on time and religious organization. . . . The Byzantine empire emerged from a fusion of a bias incidental to papyrus in relation to political organization and of parchment in relation to ecclesiastical organization. The dominance of parchment in the West gave a bias towards ecclesiastical organization which led to the introduction of paper with its bias toward political organization. With printing, paper facilitated an effective development of the venaculars and gave expression to their vitality in the growth of nationalism. The adaptability of the alphabet to large-scale machine industry became the basis of literacy, advertising, and trade. The book as a specialized product of printing and, in turn, the newspaper strengthened the position of language as a basis of nationalism. In the United States the dominance of the newspaper led to large-scale development of monopolies of communication in terms of space and implied a neglect of problems of time. . . . The bias of paper towards an emphasis on space and its monopolies of knowledge has been checked by the development of a new medium, the radio. . . . The ability to develop a system of government in which the bias of communication can be checked and an appraisal of the significance of space and time can be reached remains a problem of empire and of the Western world.[10]

As we shall see later, Innis' sense of the threat of future monopolizations of communicative power requires that we not lose sight of this issue once power shifts into the new site of biotechnology and its computerized synthesis of space and time, establishing empire over life and nature ever more deeply.

Although McLuhan enables us to grasp an intervening stage in this development, his celebration of the electronic synthesis tends to dissipate the energy needed to reconceptualize modern biopower. If in Vico foresight is farsight, then in McLuhan vision is re-Joyced into tele-vision. By way of Gutenberg, we are returned to our senses: the eye that left its body is restored to its center, a flickering omphalos. Thus, in a repetition of ancient symbolism, the modern house becomes a machine-within-a-machine whose aerial (*universalis columna quasi sus tinens omnia*) hooks it into the universe, floating our home in a Milky Way of waxes, deodorants, famines,

war, and inanity. Vico's severe poem of the world's body is now inverted—
Narcissus-like—by a world technology that communicates nothing but ourselves desiring ourselves:

> To behold, use or perceive any extension of ourselves in technological
> form is necessarily to embrace it. To listen to radio or to read the print-
> ed page is to accept these extensions of ourselves into our personal sys-
> tem and to undergo the "closure" or displacement of perception that
> follows automatically. It is this continuous embrace of our own tech-
> nology in daily use that puts in the Narcissus role of subliminal aware-
> ness and numbness in relation to these images of ourselves. By continu-
> ously embracing technologies, we relate ourselves to them as servo-
> mechanisms. That is why we must, to use them at all, serve these
> objects, these extensions of ourselves, as gods or minor religions.[11]

In the modern world our vocabularies of public and private space
and the arrangements whereby we constitute individual and collective identities are increasingly disembedded from literacy. Our private senses, like our
nationhood, have lost their closure. Indeed, if we follow McLuhan, literacy
appears only to have been a switching point in the circuitry of retribalization:

> That the abstracting or opening of closed societies is the work of the
> phonetic alphabet, and not of any other form of writing or technology,
> is one theme of *The Gutenberg Galaxy*. On the other hand, that closed
> societies are the product of speech, drum and ear technologies, brings
> us at the opening of the electronic age to the sealing of the entire human
> family into a single global tribe. And this electronic revolution is only
> less confusing for men of the open societies than the revolution of pho-
> netic literacy which stripped and streamlined the old tribal or closed
> societies.[12]

There is, however, an extraordinary falling off between the prophetic release
of *The Gutenberg Galaxy*—which McLuhan considered a footnote[13] to
Innis's concerns with the politics of communication—and his own uncritical
acceptance of what we might call the McLuhanberg Galaxy. At first sight,
we seem to be offered a more profound analysis of the structures of experience required to filter political power and its communicative media. In a
critical comment introducing Innis's *The Bias of Communication*, McLuhan
calls for the interiorization of Innis's theory of staples which would in effect
reveal how the modern state is able to implant the circuitry of power into
our very nervous system:

What Innis has failed to do . . . is to make a structural analysis of the modalities of the visual and the audible. He is merely assuming that an extension of information in space has a centralizing power regardless of the human faculty that is amplified and extended. . . . Visual technology creates a centre-margin pattern of organization whether by literacy or by industry and a price system. But electric technology is instant and omnipresent and creates multiple centres-without-margins. Visual technology whether by literacy or by industry creates nations as spatially uniform and homogeneous and connected. But electric technology creates not the nation but the tribe—not the superficial association of equals but the cohesive depth of the totally involved kinship groups. Visual technologies, whether based on papyrus or paper, foster fragmentation and specialism, armies and empires. Electric technology favours not the fragmentary but the integral, not the mechanical but the organic. It had not occurred to Innis that electricity is in effect an extension of the nervous system as a kind of global membrane.[14]

Rather than pursue the *biopolitical* issues in the bias of communication, McLuhan settled for a surrealist celebration of its commercial narcosis, scoring off realists at the expense of moralists. He thereby failed to see in television a political and commercial pacemaker implanted in *the body of desire*—not to release it—but to commit it ever more deeply to the logocentric controls of corporate and global capitalism. In this way, McLuhan abandoned the insights he once had from reading newspapers and listening to the radio, namely, that *the mechanical bride marries us to the corporate economy and to its global extravaganzas*. In such a marriage our political consciousness is reduced to a private and household amusement, inextricable from the rest of the show-and-tell that inundates us in the name of news and information.[15] In short, we lose sight of the problem of the monopoly of knowledge, as Innis called it, which is built into the administration of the media as instruments of biopower.

THE BIOTEXT: THE COMMUNICATIVE TISSUE OF POWER

I now want to show how, despite certain reservations, I nevertheless see McLuhan's thought relevant to the new contexts of biotechnology and its consequences for the body politic. To do so, I want to introduce the notion of the *biotext*, that is, the body as a communicative tissue upon which social power is inscribed, at first externally (*the sociotext*) and now perhaps from the body's very insides, if we extrapolate the possibilities of genetic editing. If this argument is at all persuasive, then we have underlined a distinctive contribution in Canadian social and political thought.

To the civilized mind, it is a mark of savagery that its people pro-duce very little else than themselves. They do not much alter their natural environment and, as it seems to us, are thereby committed to a minimal existence. We think it is a mark of civilization when the individual is severe-ly marked off from the state and the economy and even from his/her family. In this scheme of things, the individual is characterized by his/her power to negotiate exchanges, to accumulate rights and properties that exercise and consolidate a separate identity. Thus the civilized individual is horrified by the nakedness of the savage man/woman because their condition reveals that they have not acquired the power to separate the public and private realms. The naked savage is a social body, a *sociotext*. Indeed, savage societies appear to be distinct from civilized societies precisely because they write themselves, inscribe or incise themselves upon the flesh of the savage—scarifying, cicatrizing, circumcising the body that civilized men and women withhold from society with the same determination as they guard their geni-tals. Civilized man is a phallocrat, his body is his own, exposed on its own terms, a charter of freedom. It is only in his prisons, mental hospitals, and torture chambers that society still writes upon the flesh. As Kafka puts it:

> . . . Whatever commandment the prisoner has disobeyed is written upon his body by the Harrow. This prisoner, for instance—the office indicat-ed the man—"will have written on his body: HONOR THY SUPERI-ORS!" . . . Many questions were troubling the explorer, but at the sight of the prisoner he asked only: "Does he know his sentence?" "No," said the officer, eager to go on with his exposition, but the explorer inter-rupted him: "He doesn't know the sentence that has been passed on him?" "No," said the officer again, pausing as if to let the explorer elaborate his question, and then said: "There would be no point in telling him. He'll learn it on his body."[16]

McLuhan ignored the *disciplinary* or *punitive codes* that are the message in the media. However, once we do invoke this perspective, we can see that *all technology is biotechnology*. In other words, we have to begin (however briefly, as in this chapter) to analyze the various historical strate-gies whereby the living bond between the individual and society is ritual-ized (ritual is the origin of technology and the *sociotext*) and thereafter con-tinuously reproduced in historically variable secular technologies of biopower. Of course, the first technology was what Rifkin nicely calls "pyrotechnology" and this can be set off from the new "biotechnology" within which Rifkin again differentiates three further stages, that is, genetic engineering, organism designing, and the engineering of entire ecosys-tems.[17] We shall turn to these specific stages, or rather the first two, in our later analysis of the political economy of the new *bioprosthetics*. For the moment, what it is important to see is that in every case man's power over nature—or his power over life— is a power over himself (as *biotext*) inscribed through the state and the economy, and its laws and sciences (*sociotext*).

As I see it, then, all of these disciplinary strategies of power may be thought of as biotechnologies. This move is intended as a deconstructive strategy—a deliberate "misreading," if you will—whose aim is to bring biotechnology as a series of specific biological and medical engineering practices within the realm of the biopolitical. Thus we are concerned with how it is that in modern society we are devising a technology for rewriting the genetic code much as savage societies once rewrote the flesh—but in a different key, played first upon the body of desire:

> For capitalism is the stage in which all the excitations, all the pleasures and pains produced on the surface of life are inscribed, recorded, fixed, coded on the transcendent body of capital. Every pain costs something, every girl at the bar, every day off, every hangover, every pregnancy; and every pleasure is worth something. The abstract and universal body of capital fixes and codes every excitation. They are no longer, as in the bush, inscribed on the bare surface of the earth. Each subjective moment takes place as a momentary and singular pleasure and pain recorded on the vast body of capital circulating its inner fluxes, . . . in short, there is . . . a going beyond the primary process libido to the organization man. The dissolute, disintegrated savage condition, with the perverse and monstrous extension of an erotogenic surface, pursuing its surface affects, over a closed and inert, sterile body without organs, one with the earth itself—this condition is overcome, by the emergence of, the dominion of, the natural and the functional. The same body, the working body, free, sovereign, poised, whose proportion, equilibrium and ease are such that it dominates the landscape and commands itself at each moment. Mercury, Juno, Olympic ideal.[18]

The biotechnological history of the modern body is now a major research focus. It involves a simultaneous rewriting of the history of the human sciences. This is difficult to understand because social scientists are unaccustomed to dealing with *the embodied subject* whose life is at stake in their enterprise.[19] We are, of course, speaking of the human discursive productions varying from poetry to medicine, from psychoanalysis to penology, from commercial jingles to the most sacred rites of passage. Here we must focus on the historical convergence of medical discourse and the vocabularies of state and economic power which operate on the new frontier of biotechnology. Our interest, as I have said earlier, is to deconstruct our pre-conceptions of political economy and of the physical body ruled hitherto either by force or by the seductions of private desires into a public economy. On the former view, the body is recalcitrant to political and socioeconomic discipline. The constraints of society and the state, so long as they can only be enforced externally, require terrible impositions of power and discipline to make an example of the poor wretch on whose body such pain is inflicted as will inscribe in the mind of the public the law's sovereign intent.

A decisive shift occurs in the history of power once the state finds a medium of communication that enables it to exploit the connection between minds and bodies more directly than in its early theatre of cruelty. This shift occurs, as Foucault has argued, when the modern state discovers that the will to knowledge can be conscripted to rewrite the *sociotext* into the communicative tissues of life, extending biopower to every vital function of individual and collective life:

> To analyse the political investment of the body and the *microphysics of power* presupposes, therefore, that one abandons—where power is concerned—the violence-ideology opposition, the metaphor of property, the model of the contract or conquest . . . one might imagine a political 'anatomy' . . . One would be concerned with the '*body politic*,' as a set of material elements and techniques that serve as weapons, relays, *communication routes and supports for the power and knowledge relations that invest human bodies and subjugate them by turning them into objects of knowledge.*[20]

Here, then, we find a history from Innis, through McLuhan to Foucault, and work of our own, describing an *archaeology of power*, moving from the state's territorial inscription (the *sociotext*), with its theatre of cruelty, to the state's discovery of the discursive production of human knowledge, desire, intelligence, health, sexuality and sanity as a communicative network of biopower inscribed within the body, binding every body into a new Leviathan, or *biotext*. Obviously, this history cannot be told in all of its detail by any single historian or social scientist. We are engaged here in an exercise of conceptual analysis and contrast in order to mark an historical divide. Thus the modern state in its therapeutic aspect is now concerned to legislate the origins and ends of life, to contracept and to abort, to marry, separate, and divorce, to declare sane and insane, to incarcerate and to terminate life with more intensive strategies than feudal and absolute monarchies could muster. Of course, modern states also exercise power in foreign affairs, in wars and as a major component of the economy. These strategies of power are not always congruent. In liberal democracies state power simultaneously defends and undermines the mental and bodily integrity of its subjects. At its lowest points, the state now practices forms of torture equal to the horror of Kafka's penal colony. In its seemingly benign form, the modern state like the corporate economy seeks to control minds, to cajole necessary behavior into desire rather than to command it with the ultimate sanction of bioforce. In practice, the state and the economy move between these two extremes.

Increasingly, however, the therapeutic state seduces us into conformity through our desire for health, education, and employment—not to mention happiness, at least as an American aspiration. This is what I have in mind when I say *all of our technologies are biotechnologies* and that in turn they are all strategies of biopower. We wish, of course, to avoid genetic damage, and we may wish to counteract infertility or dangerous births. Our motives in this are at first humane. Yet our technologies for delivering our humanity in this respect may be inhumane. Indeed, there is already enormous concern of this score and considerable legislative activity that we cannot possibly recount here.[21] Our focus must be on how the basic metaphors of communications serve to extend *biocracy*. I do not want to exaggerate the implications of biogenetics for our political lives. Nevertheless, we should be aware that a double claim is entered in the debate on genetic engineering.[22] The first is, of course, the *technological a priori*, that is, "if it can be done, it must be done." There is, however, a rider in the second claim which brings it much closer to the first, namely that, "in science, of course, what can't be done now, may well be possible *later*." Thus the only solid objection to the technological a priori is, "even if it can be done, it shouldn't be." Here, however, the life of science, and not only of the life-sciences, is likely to be invoked as the highest conception we have of ourselves. This view is likely to prevail, I think, because we now conceive of life itself as the very elemental structure of communication (the DNA code) into which all other discursive codes can be channeled in order to amplify the expression of life.[23]

Biotechnology must presently be seen in terms of two prosthetic strategies, one now largely available, and the other increasingly possible:

1. *spare part prosthetics*
2. *genetic prosthetics*

We might think of these as two strategies for rewriting the biotext from spare-part man to self-made man.

In the mechanist vision, each organ is still only a partial and differenti-ated prosthesis: a "traditional" simulation. In the biocybernetic view it is the smallest undifferentiated element, it is each tiny cell that becomes an embryonic prosthesis of the body. It is the formula inscribed in each tiny cell that becomes the true modern prosthesis of all bodies. For if the prosthesis is ordinarily an artifact which supplants a failing organ, or the instrumental extension of a body, then the DNA molecule, which contains all the relevant information belonging to a living creature, is the prosthesis par excellence since it is going to permit the indefinite prolongation of this living being by himself—he being nothing more than the indefinite series of his cybernetic vicissitudes.[24]

The two strategies, although seemingly on the same biomedical frontier, are in fact as far apart as early and late capitalism. That is to say, the economy of spare part prosthetics involves us in a combination of med-ical craft and commercial banking and distribution procedures. Such sys-tems may be entrepreneurially or state-managed and both may draw upon voluntary donors. As Titmuss has shown in the case of blood supply,[25] there are a number of problems with quality and continuity in the supply of spare-part prosthetics. These problems could be circumvented in a number of cases, if it were possible to anticipate genetic faults and to correct them at the DNA level. Indeed, to the extent that genetic engineering is possible, we might then implant the basic market rationality of efficiency and choice at the DNA level. That is to say, we could contemplate parental choice of bio-logically perfect embryos. A mark of such perfection, from the point of view of the parent, might consist of the embryonic replication (cloning) of themselves. If that were indeed a possibility, then biotechnology would finally deliver the myth of Narcissus from its mirror. Rather, as I see it, it would defamilize the body and the imagination of future individuals making them the creature of the dominant ethos in either *the market or the state as matrix*. Under such conditions the institution of life, and not only its biocon-stitution would be radically altered. Our religious and political institutions, the Bible and Parliament, will cease to be our originary institutions. In the laboratory and the clinic life no longer has any history. Birth will become a consumer fiction like Mother's Day, and thereafter our hitherto embodied and familied histories will float in a commercial narcosis monopolized by an entrepreneurial and a statist biocracy, realizing the nightmare of 1984.

Genetic engineering is enchanting therefore because it reanimates *the myth of prosthetic man*.[26] It is all the more engaging since it appears that the biotext for this refurbished myth is encoded in the basic material of life. Even though he dismisses much of the science fiction surrounding genetic engineering, it is nevertheless interesting to see how Medawar's formulation of the historical and demographic implications of biotechnology echoes the utopian dream of the administrative state with which we have flirted ever since Plato first devised the Republic:

> At the root of all genetic engineering lies . . . the greatest scientific discovery of the twentieth century: that the chemical make-up of the compound deoxyribonucleic acid (DNA)—and in particular the order in which the four different nucleotides out of which it is assembled lie along the backbone of the molecule—encodes genetic information and is the material vehicle of the instructions by which one generation of organisms governs the development of the next. If the DNA message is altered, the effects of doing so are, in their context and of their kind, as far reaching as the effects would be of altering the wording of congressional or parliamentary legislation or the wording of telegrams conveying diplomatic exchanges between nations.[27]

Although Medawar dismisses the wildest versions of biocracy, it is significant that he, Rifkin, and Leach, whom I have quoted earlier, all consider that we are on the frontier of a potential constitutional change. For the matter is that, whether we consider our fundamental charter to lie in the Bible or in Parliament, we now envisage it being rewritten. We do so, however, with the worry that the constitutional changes involve a simultaneous rewriting of the body politic and of the politics of the body—but from an extrapolitical site which we shall explore in Chapter Nine where we consider social responses to AIDS.

NOTES

1.   John O'Neill, "On the History of the Human Senses in Vico and Marx," in his *For Marx Against Althusser* (Washington, DC: University Press of America, 1982), pp. 81-87.

2.   Jeremy Rifkin, *Algeny: A New Word—A New World* (New York: Penguin Books, 1983).

3.   *Harold A. Innis, Empire and Communications*. Revised by Mary Q. Innis. Foreword by Marshall McLuhan (Toronto: University of Toronto Press, 1972), p. 9.

4.   See Chapter Six.

5.   *The New Science of Giambattista Vico*. Trans. from the Third Edition by Thomas Goddard Bergin and Max Harold Fisch (Ithaca and London: Cornell University Press, 1970), Paras., 126-127.

6. John O'Neill, *Five Bodies: The Human Shape of Modern Society* (Ithaca: Cornell University Press, 1985).

7. Emile Durkheim and Marcel Mauss, *Primitive Classification*. Trans. and ed. with an intro. by Rodney Needham (London: Cohen and West, 1963), pp. 82-83.

8. Claude Lévi-Strauss, *The Savage Mind* (London: Weidenfeld and Nicholson, 1966).

9. *The New Science*, para. 405.

10. *Empire and Communications*, p. 170.

11. Marshall McLuhan, *Understanding Media: The Extensions of Man* (New York: McGraw Hill, 1965), p. 46.

12. Marshall McLuhan, *The Gutenberg Galaxy: The Making of Typographic Man* (Toronto: The University of Toronto Press, 1962), p. 8.

13. *Ibid.*, p. 50.

14. Harold A. Innis, *The Bias of Communication* (Toronto: The University of Toronto Press, 1951), p. xiii.

15. John O'Neill, "McLuhan's Loss of Innis-Sense," *Canadian Forum* LXI/709 (May 1981):13-15. But Kroker stresses McLuhan's analysis of the diseases of communication, see his *Technology and the Canadian Mind*, p. 73.

16. Franz Kafka, "In the Penal Colony," in *The Complete Stories*, Nahum N. Glatzer, ed. (New York: Schocken Books, 1971), pp. 144-145.

17. *Algeny*, p. 215.

18. Alphonso F. Lingis, "Savages." *Semiotext(e)* III/2 (1978):101-102.

19. John O'Neill, *The Communicative Body: Studies in Communicative Philosophy, Politics and Sociology* (Evanston, IL: Northwestern University Press, 1989).

20. Michel Foucault, *Discipline and Punish: The Birth of the Prison*, trans. by Alan Sheridan (New York: Vintage Books, 1979), p. 28, my emphasis.

21. Aubrey Milunsky and George J. Annas, Eds., *Genetics and the Law I and II* (New York: Plenum Press, 1975).

22. June Goodfield, *Playing God: Genetic Engineering and the Manipulation of Life* (London: Hutchinson and Co., 1977).

23. Gerald Leach, *The Biocrats: Implications of Medical Progress* (London: Penguin Books, 1972), pp. 153-154. For the use of other communications' metaphors such as "blue print," "punch card," "genetic alphabet," "genetic dictionary," "universal language of life," see Edward Frankel, *DNA: The Ladder of Life* (New York: McGraw Hill, 1979), Ch. 7, "DNA and the Genetic Code of Life."

24. Jean Baudrillard, *De la seduction* (Paris: Galilee, 1979), pp. 231-232.

25. Richard M. Titmuss, *The Gift Relationship: From Human Blood to Social Policy* (New York: Vintage Books, 1971).

26. Sigmund Freud, *Civilization and its Discontents* (New York: W.W. Norton and Co., 1962), pp. 38-39.

27. P.B. Medawar, "Fear and DNA," *The New York Review of Books*, October 27, 1977, p. 15.

CHAPTER EIGHT

MEDIA-culture and the Specular Functions of Ethnicity, Fashion, and Tourism

I begin with the proposition, argued in Chapter 7, that all *technology is biotechnology*. I do so in order not to lose the connection between our technological culture and the life-world which it either seductively colonizes in the name of utility and practicality or else imperiously condemns to ignorance and obsolescence. Thus from the standpoint of the life-world, technology is *antinature*. As such it has preoccupied our greatest thinkers from Marx to Husserl and Heidegger. With similar boldness, I shall consider ethnicity as the shaping element or inscape of the cultures that constitute the life-world. I do so in order not to lose the point that, whereas our technological culture is relatively universal and homogeneous, the life-world is local, ethnic, and richer in meaning and values than our technoculture.[1]

Next, I shall rephrase the relations between technology, the life world and ethnicity in terms of a classical sociological paradigm drawn from Weber and Parsons.[2] Thus we may consider a technological society to require of its members conduct governed by norms of rationality, universality, affective neutrality, and meritocratic achievement. Such a society will need to subordinate its own opposed tendencies to value nonrational, local, passionate, and kinship-governed conduct. In short, a technological society strives to subordinate kinship or ethnicity to rationality and bureaucracy, and to represent the conflict between technology and ethnicity as a problem of *social control* requiring the imposition of homogeneity over difference. As we shall see, the same imperatives rule with regard to sexuality and the crisis of AIDS which we examine in the following chapter.

Finally, in order to provide a key to the apparent arbitrariness of my use of the tourist, Kojak and Third-World fashion as *figurations of ethnicity* in the technological age, I shall translate my argument a third time into the language of natural symbols, as developed by Mary Douglas.[3] The effect I want to achieve is to show how technical rationality and ethnicity may be considered figures of opposition linked to the contrastive structures of mind/body, culture/nature, order/disorder. These terms are inextricably bound, however much we try to give hegemony to the symbols of techno-culture. Our civilization is neurotic about its body, its family, and its ethnicity, inasmuch as these become figures of disorder in an age whose faith lies in the smooth operation of a technobureaucracy far removed from such local impurities:

> According to the rule of distance from physiological origin (or *the purity rule*) the more the social situation exerts pressure on persons involved in it, the more the social demand for conformity tends to be expressed by a demand for physical control. Bodily processes are more ignored and more firmly set outside the social discourse, the more the latter is important. A natural way of investing a social occasion with dignity is to hide organic processes. Thus social distance tends to be expressed in distance from physiological origins and vise versa.[4]

By reconstructing the technology/ethnicity relation in terms of the "purity rule," we can understand how technological societies treat ethnicity as "difference," or as "other" to the homogenizing processes of rationalization and bureaucratisation that attempt to exclude all other forms of the life-world as alien modes of heterogeneity. Thus the ethnic, the native, the family, the feminine, the exotic, the wilderness, the virus become *natural symbols* opposed to the *technosymbols* of the hegemonic culture of industry, science, and technology. In practice this contradiction is compromised. As I shall show, the technoculture strips ethnicity in order to reappropriate it in a secondary system of natural symbols floated in the *social imaginary* whose specular production is the primary function of the media. Tourism, crime, and Third-World fashion can then be seen as figurations of ethnicity operating as the internalized/externalized otherness of the technological world order and its own desired specularization.

It is, of course, impractical to unthink our technology. What can we mean by such an exercise, what use, what good can come of it? Such questions themselves reveal the extent to which we have made technology our voice, our omniscient and ubiquitous sensorium, our pride, our morality. Indeed, we have no other point of view than the view we have of ourselves amplified, and relayed through our specular technologies. Nor have we any history, society, politics, or economy outside of our video terminals which now reproduce us in light matrices that owe nothing to the womb, and less to our ordinary and divine memory. For the flash which places us at the center of the screen renders us the outcasts of mankind's hitherto civilized incorporation and continuity.

From the standpoint of the life-world, it is becoming apparent to us that the *technological age is an age of death*, an age whose end obliges us to revision its origins. In short, our fire technology no longer inspires us with its Promethean revolt and even its domestic achievements have degenerated into the endless gadgetry of obsolescence and mindless consumerism. Whereas our technology once promised to make us lords of creation, we find ourselves cosmic aliens, polluters of nature, and barbarians of the galaxy. Our gods are the twisted creatures of that cyberspace which in our own inner space we no longer revere. Our ambition is to travel to the stars in order to extinguish them in the darkness our planet now casts upon the galaxy. Today's children know this lesson while their elders continue to run away from it. Meantime our children grow older than their parents in a world whose seductions deepen the exploitation of their minds and bodies, driving them into helpless narcissism and suicide on their own doorstep, while elsewhere condemning them to famine and genocide, short of musical relief.

To argue that we are *dispossessed* and are *disembodied* in the technological age, will surely seem perverse, if not fantastic. Everything—every machine—seems to tell us the opposite. For all our things and all our machines are talkative. They are the principal rhetorical agents for the view spectacularized in our media—in everything as a medium for the media—that we are in charge of things and of ourselves. In such a world it is the machine-less, the information-less that are the dispossessed, barely conceivable figments of history and nature before the age of the Coca-Cola bottle, before the gods had gone crazy, before writing and the American Express Card had permitted everyone to leave home. Thus all shrines lead to our technology and all our pilgrimages honor our ability to travel the earth and to compare its ethnic peoples in the image of history's most homeless citizen—the Rambo tourist. Tourists, of course, do not lack those practical arts which render them, however single or infirm, however ignorant or indigent, the temporary lords and ladies of the vacationer's empire. The tourist honors the world technological order. Indeed, *the tourist is the world's inspector general*, roaming freely in a Disney World of ethnic food, drinks, and crafts arrayed for his pleasure and profit.

Nothing can be more curious than the annual flocking of American immigrants to their former homelands to discover the now omnipresent toilet roll, toothpaste, central heating, hamburger, and Coca-Cola signs for which their ancestors displaced themselves and with which they identify their choice of America as the ultimate trip. What is sad in all this is that the real history of immigrant labor and political refugees recedes into the back ground of *canned cultures* created for the international tourist who travels as nearly as possible within the same technosphere as he or she enjoys at home. Today's tourists now spread around the world in a few hours with minimum discomfort and colossal self-confidence in the relative value of their national currency. Until recently, they were by and large a good-natured people willing to suffer strikes that often leave them stranded before or after a holiday, not to mention the occasional hijacking and severe bouts of diarrhoea. Indeed, it once seemed that tourists could only be angered by those natives who short-changed them in a deal where their own honor as the cunning collectors of worthless things was at stake. Today, however, tourists are outraged to discover that they are pawns in a techno-order that has colonized the world pushing it into a picture book past (or the more insulting scenarios of the lost American Express card) which "unaccountably" explodes and rages against our obscene cameras. Thus *Rambo-tourism* now meets its counterpart in terrorism which threatens to limit that thought-less empire where travel narrows the mind.

Have we misplaced our observations upon the tourists, hitting at those innocents abroad rather than looking for their booking agents? Indeed, it might well appear that we have chosen sarcasm in place of criticism, enjoyed irony where we might have suffered the more laborious tasks of analyzing the political economy of multinational corporatism and its hegemony over the earth and interstellar travel. Such an analysis is necessary and we will turn to it in our closing pages. At this point we wish to add something that is just as difficult to capture because it does not seem to be part of the puzzle. The part we are trying to fit to the puzzle is the phenomenon of the *specular ideology of technology*, or rather, the practices through which we moralize the behavioral, mental, and social commitments which underwrite our technological age and its specularization as an age of benign utilitarianism. The ideological production of the specular image of the technological age operates through the figuration of the citizen, the deviant, the consumer, the child, the tourist, the patient, the feminist, the ethnic, the handicap, and the transplant as celebrants of the social system that reproduces their identity and its supporting socioprosthetics.

We speak of the *floatation* of these figures because, although the fundamental drive of technological society is to subordinate all social relations of production to the basic operations of the forces of production as narrowly conceived in the laws of profit and practicality, on the cultural level the subordination of the social to the technical infrastructure is reversed, or floated in ideological discourses that figurate the individuals as the principal agent of social processes. How is it possible for the work of justice, for the hard police work of crime hunting, to be achieved by an ethnic in a city where "everyone knows" crime is produced and not solved by ethnics? How is the impartial, rational, legal-scientific work of the bureaucratized criminal system executed by ethnics whose values are inimical to such practices? Part of the answer is that in reality the police solve very little crime. To reduce crime in any serious way we need massive reforms in the property system and its effects upon levels of education, housing, employment, urbanization, race, and ethnicity. What a capitalist criminal system needs to portray in its daily operations is the relatively honest efforts of the lower classes *to police themselves* by repressing ethnicity as the source of its troubles. This is the basic specular function of all police and detective series (choose your favorite) that now spread from crime to law to medicine.

Because its expressive task is primary, the police movie needs to be guaranteed a modicum of success at the instrumental but secondary level of actual police work. The large amount of specular work devoted to expressive ethnicity is therefore redeemed through a benign technology (bugging and informing) which provides for the hunted to tell Kojak (or Anycop) where to hunt for them. Here, then, are the two sides of the moral economy of police work. Because of an unfailing technology of disclosure (however little it transforms the field of crime), police work can be made to float all the other expressive values of the society which it represents as a moral order. Thus in the pastel world of *Miami Vice*, which avoids the earth tones of ethnicity in favour of the electricity, America's doubly failed ethnic adventures at home and in Vietnam are refloated in the black/white police couple, Ricardo Tubbs and Sonny Crockett.[5] Here the specularization of police work—which costs millions of dollars per show—exceeds the cost of much real police work. But since police work is not meant to solve the drug problem, its specularized functions are again primary and make it compatible with elaborate body advertisements for fast cars, yachts, Italian tailoring, and the throb of ethnicity. Because the American Dream remains an exclusionary vision for the lower orders, the city is the perfect setting for the morality plays serialized in *Kojak*, *Police Woman*, and *Miami Vice,* and so on. This is why *Miami Vice* went to such trouble to color its world, remaking the cold city. Thus, to repeat an earlier observation, the urban environment, and in particular the ghetto, is the prime target of effacement. Everything that is contradictory and incoherent in the material basis of the political economy is reflected in the neonized iconography of the urban environment, in its wealth and poverty, its comforts and dangers, its crime, its sophistication, and its vulgarity. The city thereby furnishes the ideal setting for a genre of television series, from *Kojak* to *Police Woman*, that serves to spectacularize the dream of *law and order as an urban morality play*, hence the current attraction of the series *Law and Order*.

Is there any connection between the figures of the Rambo-tourist and Anycop, and how can they reveal the life-world of the technological age? The question might be repeated to ask how these two figures give us any insight into the figure of ethnicity as the internal "other," or the external "other" of the technological society and the "world order" it seeks to impose upon the life-world. The connection I am suggesting lies in the operations of the *specular technology*, or media of the technological society, which display it as a moral order which needs the *Pax Americana* at home and abroad. Thus it is now possible for the United States to treat terrorist attacks upon its military and diplomatic operations as attacks upon the outposts of American tourism. The American tourist is in turn capable of rebutting Italian protests against a McDonald's fast food pit adjacent to the Piazza d'Espagna with the rejoinder: "I have a right to have a Big Mac wherever and whenever I want to." This right to the great American fry is ultimately backed by firebombs, napalm, and the nuclear fry. Such is the moral order of the things we consume without any thought for how it is we are dispossessed by the "lifestyle" with which they threaten the life-world. Meantime, while Americans play sheriff to the world technological order, at home one of their own terrorists turned a local McDonald's into a massacre reminiscent of Mai Lai. Worse still for the faceless family restaurant, the Chicanos turned it into a shrine, making memories somewhere that was forgettably nowhere.[6]

We have now to draw together the way techno-symbolism and natural symbolism work together in the specularization of ethnicity as the underside of the world industrial order. We propose to do this by treating Third-World fashion as a biotechnological integrator of the civil and savage bodies of international capitalism. We consider fashion a body technology inasmuch as it commits us to homogenized styles, cut, and colors which rule the seasons with an imposed "look," while simultaneously invoking the body's freedom and frivolity in the choice offered between the elements that compose the "look." To achieve this, *the fashionable body* must learn to be alienated from itself through its desire to become itself by abandoning previous commitments to styles that always threaten to be out of style. Fashion, then, is a technique for the creation of comfortable risk and boredom through endless flights from the stylized body whose labor is freely engaged on behalf of the fashion industry. The world economy of fashion is in turn deeply committed to the ruthless exploitation of sweated labor, as well as the pillage of its cottons, silks, colors and designs.[7] The expropriation of Third World fabrics, colors and designs produces the natural symbols of the preindustrialized body. Floated in the post-industrial semiurgy of classless, raceless, timeless, lifestyles, the fashionable body of late capitalism denies the domination of its own internal and external ethnicity. At the same time, it entertains its repressed fears of the exotic cultures of poverty and colonialism symbolized in the Bazaar/bizarre world of aestheticized exploitation.

The operation of this aspect of the logic of late capitalist semiotics is nicely caught in Julia Emberley's remarks on the text and imagery of John Galliano's spring collection in *Harper's/Queen* magazine for February 1985. While seemingly far away from the pastel world of *Miami Vice*, Galliano's "Visions of Afghanistan: Layers of Suiting, Shirting and Dried-Blood Tones" in fact conceals the same violence, the same genocide, the same bloody bodies at the end of the industrial violence of the worlds two major powers. What Emberley discloses is the double layering of textual and visual signs required to appropriate the broken symbols of colonized ethnicity in the forced alliance of tradition and fashion:

> In the syntax of Galliano's title we find the heterotopia, a heterogeneous splitting and fracturing which is translated in the "world of fashion" as a multiple and spectacular field of types and tropes that circulate on the surface of visual and textual representations. The fashion-effect of his title dismantles the narrative continuity of presentation because its syntax is broken, dismembered, shattered and replaced by a "layered effect"—a horizontal syntax, discontinuous and fragmented, gives way to a vertical effect of imaginary and semantic layers. In fashion, images cut across traditional barriers or limits of representation, effacing along the way differences and historical specificities and producing, instead, a unitary effect of congenital pluralities that apparently "hold together" without contradictions.[8]

Thus, as we observed in the beginning, the lifestyles of the technological age turn out to be modes of antinature, styles of death that live off a degraded life-world pushed into the urban ghetto and the colonial slum. Third World chic, punk and pastel are merely the death masks of late capitalist history whose narrative fragments into the discontinuous lights and deafening sounds in which it records itself.[9] In turn, the layered poverty, and masqued pallor of its unkempt youth turns against the world's sick and poor its own mirror-image, embracing the world's children in the profitable protest of a charity that never gives away the game.

Such observations are difficult because they imply standards of reason and justice that reject the postmodern insistence upon the pluralization of culture. The latter confuses center and periphery in the development of global capitalism. What I mean is that the plurality of cultures is available to us only as a result of the marginalization and fragmentation of local cultures in the face of the globalization and homogenization of late capitalist corporate culture. Curiously enough, the melancholia and electicism of the postmodern plurality saps its individual producers and consumers precisely because they have lost all dimension of oppositional culture—of ancient, distant, exotic, class, ethnic, and gender cultures. Within national cultures, the deconstruction of the Law, of patriarchy, and the phallus are therefore attacks upon the abandoned fortresses of late capitalism which is everywhere and nowhere. In fact, by allowing its loyal opposition to attack its presumed notions of authority, art, sexuality, and politics, late capitalism achieves a benign solidity and tolerance that in turn underwrites its postmodern university departments where such criticism flourishes.[10]

Thus postmodernism and feminism are necessary correlates since their celebration of the autonomy of signifiers naturally admits the expanded circulation of the now least valuable social signifier—(woman)—into its orbit. We are referring, of course, to the reduction of the local, historical, cosmological value of woman's difference to her complete exchangeability in markets whose own discursive production is the feature that best exemplifies the direction of late capitalism and its recoding of the life world and its human shapes.[11] This is not to say that women are any less wise than any other minority demanding an equal opportunity to become unequal within a social system they do not otherwise challenge. Such an insistence, as Marx would have observed, merely deepens the levels of surplus value extractable from the free market operation of late capitalism which at the same time yields to various welfare and therapeutic demands as palliatives to the negative side effects of its overall operation.[12] What is peculiar to the feminist narrative is that it is floated in an imaginary universe of symbolic equality which forecloses on its own signified domination of those men, women, and children who live with the underside of success in executive suites, panels, committees, galleries, and classrooms. Thus the male gaze is not deconstructed by its female return so much as by its very absence from a world in which the human sensorium is subordinated to the look of things, or to the light-sound of events which constitute the disembodying information process of late capitalism.

Neither mastery nor victimage can be espoused in a cultural system that can recycle all of its class, sexual, artistic, and political symbols to re-embody a-temporally and a-spatially configurations whose social contexts no longer delimit the places of late capitalism. Thus a woman on the move, selling other people a move, can wear a perfume celebrating the flight of the Vietnamese peasants from a murderous machine-gun fire that destroyed their villages and countryside, threatening to turn their land into a great American parking lot. Even this horror, captured in the photographs of burning women and children, can be expropriated as a challenge to American sentimentalism and its medicalized charity. The media exposure of the Vietnamese mother is due to an imperialist system of pillage whose genocidal impulses put Third World women well beyond the recycled emancipation of women in the advanced industrial countries. To speak of solidarity under such circumstances is merely to bring Third World women under the gaze of women within the hegemonic world. Thus to speak of visibility as a male prerogative, or as a phallocentric practice which produces women's invisibility for herself, merely distracts from the invisibility of the socioeconomic system which reproduces these and similar practices while leaving itself without a name. To locate these practices in patriarchy when in fact the majority of men, no less than women and children, are powerless is a self-inflicted injury that further incapacitates families whose lack of authority marginalizes them and their political function.

In short, to *sexualize the system of stratification and imperialism rather than to analyze how that system is productive of racist, sexist, and class ideologies is really to be mesmerized by the surfaces of exploitation.* Worse still, it is to risk cooperation through inclusion in the imagery effects of participation and visibility which is the prime effect of what I have called the specular ideology of the technological society and its effacement of ethnicity, family, gender, and locality.

NOTES

1.   John O'Neill, *The Communicative Body: Studies in Communicative Philosophy, Politics and Sociology* (Evanston, IL: Northwestern University Press, 1989).

2.   Talcott Parsons, *The Social System* (New York: The Free Press of Glencoe, 1951).

3.   Mary Douglas, *Purity and Danger: An Analysis of Concepts of Pollution and Taboo* (London: Penguin Books, 1970); *idem, Natural Symbols: Explorations in Cosmology* (London: Penguin Books, 1973).

4.   *Natural Symbols*, p. 12.

5.   For alternative interpretations of *Miami Vice*, see Todd Gitlin, "Car Commercials and Miami Vice: 'We Build Excitement'," in *Watching Television*, Todd Gitlin, ed. (New York: Pantheon Books, 1987), pp. 136-161; and Steven Best and Douglas Kellner, "(Re)Watching Television: Notes Toward A Political Criticism," *Diacritics* (Summer 1987):97-113.

6. John O'Neill, "McTopia: Eating Time," in *Utopias and the Millennium*, edited by Krishan Kumar and Stephen Bann (London: Reaktion Books, 1993), pp. 129-137.

7. Rosalind Coward, *Female Desire: Women's Sexuality Today* (London: Granada Publishing, 1984).

8. Julia Emberley, "The Fashion Apparatus and the Deconstruction of Postmodern Subjectivity," in *Body Invaders: Panic Sex in America*, Arthur and Marilouise Kroker, eds. (Montreal: New World Perspectives, 1987), pp. 47-60.

9. Dick Hebdige, *Subculture: The Meaning of Style* (London: Methuen, 1979).

10. John O'Neill, *The Poverty of Postmodernism* (London: Routledge).

11. Ivan Illich, *Gender* (New York: Pantheon Books, 1982).

12. John O'Neill, "The Disciplinary Society: From Weber to Foucault," *The British Journal of Sociology* XXXVII/1 (March 1986):42-60.

CHAPTER NINE

Media AIDS: Communicable Disease, Sexual Ideology and Global Panic

The 1980s will be remembered among other things as a decade in which a series of human troubles assumed "global" proportions. These troubles, ranging from the punctured ozone layer to AIDS, owe much of the attention they receive to the global communications system through which they are reported. At the same time, however, they must be regarded as elements in the creation of the very global order in which they are reported as "threats," "progress," or "acts of God" that in turn require the energy and concern of global institutions ranging from the World Health Organization to air lifts and rock concerts. Many of the world's epidemics are now, so to speak, *media plagues* that reach global populations regardless of their immediate political and socioeconomic context. In the case of AIDS much of what the world knows is what the media makes of it. *Media-AIDS* affects everyone regardless of their chances of contracting the HIV virus. *AIDS is the communicable disease* and as such it creates the greatest dis-ease in a communications society already stressed and rendered violent by its high-stress media.

This does not mean that HIV/AIDS is merely a mediatized illness, or a journalistic plague. It means that we are obliged to analyze it as a challenge to both our biomedical knowledge and our knowledge of communicative processes in advanced societies that impinge severely upon the rest of the world. At the same time, such analysis is necessary for our own sake because it is our society which shapes the discursive practices of our understanding of AIDS and of the sort of society we value in response to its troubles:

> For AIDS, where meanings are overwhelming in their sheer volume and often explicitly linked to extreme political agendas, we do not know whose meaning will become "the official story." We need an epidemiology of signification—a comprehensive mapping and analysis of these multiple meanings—to form the basis for an official definition that will in turn constitute the policies, regulations, rules, and practices that will govern our behavior for sometime to come . . . these may rest upon "facts," which in turn may rest upon . . . deeply entrenched cultural narratives.[1]

Having, like everyone else, spent hours on the media's coverage of AIDS and its soap-opera attempts to educate the public through "awareness" of AIDS, I shall try to locate this new global phenomenon in terms of the ideological order to which I think it belongs and to which it offers such a threat. This is a difficult exercise since it is always hard to turn one's eyes away from the screen in order to see what the show might be about, after all.

In the present context of HIV/AIDS knowledge, marked as it is by the continued absence of a vaccine discovery, it falls to the social and health sciences broadly conceived to devise institutional responses to AIDS that will contain both the illness and our social responses to it. Here, of course, "containment" is sought on two levels and it may be that a considerable fiction is involved in hoping that the HIV virus is, so to speak, *virus sociologicus*. While it is doubtful that the virus can learn sociology, it is certainly true that sociology cannot remain ideologically ignorant of virology. But this in turn means that the social sciences in general are obliged to rethink themselves before they can be adapted ready-made to the new limitations which HIV/AIDS imposes, not only upon our sexual conduct, but upon a range of professional behavior where contact with AIDS is involved. This is especially the case since the professionalization of the social sciences, like that of the medical sciences, has proceeded in terms of an ideological demarcation between factual knowledge and moral knowledge that, while honored by hardline scientists of either ilk, is in fact breached by developments in the biotechnological sciences that have reopened the frontier of ethicolegal inquiry. The social sciences no longer have any neutral ground in these matters, and this is particularly the case with those who suffer from AIDS since they oblige us to reconceive our social policies and our moral values regarding trust and community.

Although AIDS is a relatively recent phenomenon, and despite the complexity of the virological and epidemiological dimensions of the HIV virus, it may be said that we have acquired a considerable knowledge of it in the short span of little over five years research.[2] Indeed, AIDS has moved quickly through a cycle of first stage of relative ignorance, followed by intensive research and discovery, to a plateau where we are waiting for the breakthrough which would permit us to counteract the HIV virus by means of a vaccine. This state of affairs has only been achieved through a considerable pace of biomedical research[3] and the involvement of governmental health agencies from the federal to municipal level, including hospitals, clinics, gay community organizations, AIDS workshops, hotlines and pamphleteering, in addition to a constant reportage of AIDS information in newspapers and on television. AIDS has even generated its own art forms in theater, film, and folk art.[4] All these activities, then, have combated social ignorance with social awareness. As a new stage of conscience-raising, AIDS awareness is now projected to recruit even preadolescent children. The cardinal virtues of contemporary sexual citizenship are exemplified in our awareness of the practices of contraception and abortion, as well as the achievement of copulation and the avoidance of AIDS. It should be observed that by the same token a great deal of *secular faith* is involved in the assumption that "awareness"—which varies from sect-like membership to glancing through myriad minor pamphlets—will alter sexual conduct to any great extent. "Safe" sex is not so easily institutionalized if only because the concept of "sex" is itself not to be understood outside of any extraordinary range of social behavior where "excitement" may preclude "safety" in any form.

I want to argue that, despite what we already know about HIV and AIDS, any further development in our knowledge and the pedagogies to be devised in public education programs is confronted with the phenomenon I shall call *socially structured carnal ignorance.* Here, what I have in mind is a number of factors that determine bodily conducts as necessarily, willfully, and desirably matters of "unknowing," of "spontaneity," of "passion," of "desire," or of "fun" and "fantasy." In short, we use our clothing, eating, drinking and sexual behavior to achieve relationships and end-experiences that may be considered moral or immoral, rational or irrational, competitive or communal, safe or risky. It is not our task to pursue the cultural, class, and gender and age variations that operate here,[5] although these aggravate the phenomenon to which I am referring. Nor have we to recount the historical and ideological shifts in the codes that dress our bodily conduct, provided we do not overlook them. It must suffice to articulate with regard to

AIDS the following elements that condition a structure of ignorance which in turn generates the fear of AIDS upon which so much public energy is expended:

1. The carrier may not be known to him/herself due to the latency period of up to 8 years;
2. The range of risk behaviors may not be known to potential victims;
3. The HIV carrier may not be known to him/herself;
4. The pursuit of epidemiological knowledge regarding the HIV virus may conflict with the civil rights affording "nonknowledge" of persons' behavior and associations;
5. The code of civil rights may guarantee nonknowledge of persons in ways that prevent or conflict with medical, police, educational, employment practices of testing and identification of persons with HIV.

In turn, within the general population, carnal ignorance may be valued in such experiences and settings as:

a. sex	d. at parties
b. drugs	e. at concerts, clubs
c. alcohol	f. on the street, in alleys, in cars

The specific behavioral codes defining these experiences where reason "goes on holiday," so to speak, will vary according to membership in a variety of settings and practices shaped by:

a. age group
b. sexual ideology, e.g., consumptive rather than reproductive sex
c. sexual identification, i.e., homosexual, lesbian, bisexual and/or heterosexual
d. religious beliefs
e. ethnicity
f. socioeconomic class, i.e., level of education, income and professional ideology.

There exists an enormous bias in the social sciences against the study of the ways in which our ignorance, misinformation and deception is socially structured. For this reason our enlightenment with regard to sexual behavior proceeds much more slowly than the deliverance promised to us by our present rationalist bias towards knowledge. If we are to make any progress in devising sexual pedagogies to respond to sexually transmitted diseases (STDs), specific empirical and ethnographic data on the social structure of carnal ignorance, as I have outlined it, must be gathered.[6] In part, such information is already available in material collected from the standpoint of the sociology of attitudes, beliefs, and opinions. What can be said, however, is that those who are least educated and most socioeconomically underprivileged in virtue of age, class, racial, and colonial status, will bear the brunt of prostitution, drugs, and HIV-infected births in the urban areas of the world which constitute a "fourth world" of social problems wherever they are situated.

With these remarks in mind, I shall turn now to the development of a framework for the study of AIDS which sets national and local concerns in the context of the *global political economy.*[7] From this standpoint, AIDS must be considered as one of a number of panics of a political, economic, financial, and "natural" sort to which the global order responds with varying strategies of crusade, sentimentality, or force. By a *globalizing panic* I understand any practice that traverses the world to reduce the world and its cultural diversity to the generics of Coca-Cola, tourism, foreign aid, medical aid, military defense posts, tourism, fashion, and the international money markets.[8] Since these practices are never quite stabilized, their dynamics included deglobalizing tendencies which will be reinscribed by the global system as threats to the "world order." Some nations may consider themselves to be the prime agents in this world order, while others can only maintain an aligned status, or else are allowed to enjoy a toy nationality, like that of Canada, that can be appealed to in order to supply neutrality functions on behalf of the world order. The globalizing panics that confirm the world order rely heavily upon the media and television, newspapers, magazines, films, and documentaries to specularize the incorporation of all societies in a single global system designed to overcome all internal division, if not to expand into an intergalactic empire. Such a vision is confirmed, for example, by the performance of chemical experiments under the gravity-free conditions of space flight which may enhance the future discovery of an AIDS vaccine. Thus, sexual practices that would not be tolerated within the social system of the space capsule, nevertheless, provide ideological justification for global medicine's quest for a perfect experimental environment. Furthermore, the relatively ghettoized sexuality of gays and blacks and Hispanic IV drug users finds its projection in the starry heaven of the heroes of all-American science, war, and medicine. Meanwhile, this medicine is largely unavailable to the poor in the United States and in the so-called Third World whose infants are ravaged by disease and death amidst populations that are continuously uprooted by famine, flood, and warfare. By the same token, media images from this part of the world are exploited on behalf of the promise of the global order whose own political economy is largely responsible for the so-called "natural" disasters that ravage the Third World, notwithstanding first-world aid.

The sexual economy, which must be treated as the framework for any grasp of the political economy of AIDS, is subject to every other sub-system of the global economy and national political economy. It is so when it appears most disengaged because its disengagement celebrates the processes of disenfranchisement elsewhere in the society, that is, the degradation of gendered economies, of family, and of church authority, as well as of any politics grounded in these communities. *When the general will is sexualized, politics are privatized; when politics are privatized, the general will is idiotized.* The politics of desire are the desire politics of a global economy which entirely escapes articulation in the speechlessness of sexualized desire. Meanwhile in the West, the postmodern insistence on cultural fragmentation implodes all differences and reduces everything to shifts in fashion and constant revision of the spatio-temporal order of global capitalism, which itself remains class and colonial in nature and cannot be thought of in terms of the shibboleths of sexism, racism, or anti-humanism, which it has outlawed at the same that it is the principle source of these very phenomena, because reformism on these issues is anything but revolutionary.

It might be argued that postmodernism celebrates a fantasmatic economy of sex and power while remaining tied to the market for global rock, drugs, and fashion and to the concert politics whose evanescent sentimentality reflects and deepens the global exploitation.[9] From this standpoint, postmodernism is the "classless" culture of a globalizing economy that exports its industrial basis wherever labor is cheapest while dividing its internal economy into rich and poor service sectors. In these sectors activities are neither community nor self-building. Here the *political economy of the signifier without sign* (family, class, race, gender) is operative. Since there is nothing at the center of a doughnut or of a dress, variety is everything. Since money has no absolute value, variety is the only norm. In this culture, those who look for signs are traditionalists, semantic fools, or semiotic idiotics—they are out of style, out of touch. Nothing looks worse on television than nationalism, fundamentalism, and anti-colonialism with their murders, their starving children, their destruction and immolation. The same is true of the images of domestic poverty, urban decay, illiteracy, and alienation that are floated without any attempt to implicate the class system. To be "it" is to be "out" in the game of global circulation.

The global economy is concerned equally with the promotion of individualism and its sexualized erasure, as we have seen in earlier chapters. Its primary politics are those of corporate identity which, in turn, conscripts an aggressive individuality pitted against his or her own nature and community. The active, young, calculating, realistic, and hedonistic recruits to corporate capitalism and its cosmopolitan culture are the idols of global media culture. They move like gods and goddesses amidst the debris of urban crime and desolation, cocaine colonialism and the life and death struggles of high-tech medicine, war, and "communication." To this end, global culture is perfectly "uni-sexual," that is, it is a same-sex culture whose technological infrastructure is indifferent, benign, or emancipated with respect to its male and female protagonists.[10] In turn, this monoculture refigurates itself as "different" by means of its agonistic pursuits of profit, sex, drugs, peace, health, justice, and progress. Here winners are "hyper-men," a caste that includes "executive" women who have proven they can be winners within the agonistic culture of global high technology and corporate capitalism. That is why it is important that females not be excluded from the business and social science professions and especially that they not be excluded from the police and military forces. Global capitalism is "unisexual" and it offers every prosthetic and therapeutic aid to the monobodies required to service its way of life. Against this promise, AIDS may be understood to constitute a potential global panic on two fronts; namely:

1. A *crisis of cultural legitimation* on the level of global unisexual culture;
2. A *crisis of opportunity* in the therapeutic apparatus of the welfare state and the international medical order.

AIDS threatens to produce a *crisis of cultural legitimation* because it tempts highly committed individuals to withdraw from the unisexual culture of global capitalism and to renounce its specific ideologies that

1. Sex is the most intimate expression of freedom and choice in the market society;
2. Sex is consumptive and not necessarily productive;
3. Sex is genderless, unfamilied, classless, and homeless;
4. Sexual repairs are available through the biomedical and psychiatric services of the therapeutic state;
5. All high-risk behavior on behalf of global capitalism will be supported by its prophylactic and prosthetic technologies whose ultimate aim is to immunize its members against the adverse environment of their own risk behavior.

The experience of AIDS panics the sexual culture of global capitalism in several ways. In the first place, it has "disappointed" those who were most committed to its ideology of sexual freedom. To its credit, the gay community has learned that its sexuality cannot be played out in the anonymous intimacy and extraordinarily high rates of casual contacts that were enjoyed in the bath houses. This reevaluation has taken effect because gays constituted a quasi-community marked by levels of literacy and organizational skills beyond what can be found in the IV drug alleys.[11] However, "heterosexuals" have been tempted to turn against the gay community in a number of ways that threaten to degrade the civil rights achievable in capitalist democracies

Business, educational, and medical institutions have all been strained by the fear of AIDS. At the heart of these delegitimizing strains in the social order is the continued lack of any vaccine against HIV which would, as it were, immunize our society against its own responses to what I shall call *AFRAIDS*. Short of a vaccine, AIDS constitutes a panic both on the individual life level and on a collective level where AFRAIDS threatens to undermine the order of civil liberties. To the extent that panic spreads, especially where AFRAIDS or the collective fear of AIDS generates secondary fundamentalist and revisionist panics, the sexual economy of global capitalism is threatened with a crisis of legitimation. AFRAIDS, however, strikes most deeply into the legitimation process when it prompts the general population in a rationalized industrial society to question the probability value of scientific knowledge with demands for absolute certainty or for the immediate availability of drugs, such as AZT, where the experimental controls and clinical tests properly required may be short-circuited by the clamor for immediate reduction of suffering, "fast-tracking" hope despite the necessarily cautious pace of research. Here, modern medical knowledge is particularly exposed since rational experiment, placebo practices, and the ethicolegal issues in patient consent reveal the costs as well as the benefits of our commitment to the industrial order. This order is threatened less by proposals to ghettoize AIDS patients, however politically crude such a suggestion may be, than by any loss of commitment to the protocols of the medicalization of health and happiness.[12] As things stand, there is a considerable need to attempt some realignment between community-based medicine, corporate research, and national medical research.

At the very worst, AFRAIDS threatens the liberal order of global capitalism with a "rebarbarization" of its social bond. Hitherto, it was possible to sustain the global fantasy of a social order without deep commitment, as in the figure of the American Express man whose creditability lies in his credit and whose faith lies in the fiction of a card that would make him at home in the homeless world of global capitalism. Will the American Express Card guarantee immunity against foreign HIV tests, and even if it did, can we be sure that American Express will issue its precious cards at home to persons without a prior HIV test? Or will those who carry the American Express Card, having passed an HIV test, constitute a new biological order of eligible capitalists? However bizarre such questions may seem, they already have their counterpart in the experiment with AIDS-free singles clubs. What is extraordinary in such responses is that they threaten to return the capitalist order to a *purity rule*, that is, to return us to a social order founded upon a *contagion model* of social relations rather than on the present *contract model* of society that has been the engine of our extraordinarily globalizing history. If this were the ultimate consequence of the AIDS panic, global capitalism would have succeeded in rebarbarizing itself due to an unfortunate contingency in its sexual culture, rather than due to its other fantasized threats of interstellar barbarism.

AIDS presents *a crisis of opportunity* in the global culture of late capitalism precisely because its immediate features, namely:

1. Autotoxicity
2. Fatality
3. Absence of vaccine

renew demands upon the therapeutic state and its biomedical apparatus to provide interim care, sociolegal immunity, and a vaccine. To do this, the social and health sciences have already been recruited to furnish ethical, economistic, and pedagogical discourses, conferences, pamphlets, multidisciplinary research, and media treatments to the public. Here, the supply of AIDS "knowledge" and pedagogy to the public whose fears demand it creates a perfect symbiosis between State Power as Knowledge, on the one hand, and knowledge as individual power/ powerlessness, on the other hand. Individuals attempt to learn that only they can stop AIDS, or Drinking/Driving, or Waste, whereas these are cultural complexes produced by and (re)productive of the ideological and therapeutic culture in which they live. In this process individuals learn that

1. The State is the ultimate producer of knowledge/science;
2. The State is the ultimate producer of health, education, and employment;
3. The State is the ultimate guarantor of civil liberties;
4. The State is the ultimate guarantor of all ideologies;
5. The State is the ultimate producer of the State and of Society and of the Individual.

It is important to see that whatever ways we may yet devise to speak about our sexuality, sexual disease, sexual discrimination, sexual liberation, our discourse is shaped by a pre-existing field of institutionalized discourses that have been authorized by the church, the state, and the sociomedical sciences with a concern for public and private welfare. These official discourses contain both prescriptive and descriptive technologies to which different professional groups claim legitimate access and application in the complex of power-knowledge-pleasure that may be called the "therapeutic state." Thus, the HIV/AIDS phenomenon enters into a highly structured field of discourse of social policy, health, and moral ideology that is contested by church, state, family, school, and secular countercultural groups with varying degrees of progressive and fundamentalist beliefs in the possibility of social control of youth, sexuality, drugs, and family breakdown.[13]

Medical "AID" might have functioned as a pretext for a social backlash against the civil rights of gays, lesbians, and the achievement of gender equality to the extent that they were seen as the source of troubles plaguing groups marginalized by the larger global economy in which these movements function. For want of a comparable pedagogic effort from the national center, these marginalized groups are appealed to by televangelists who adopt an extraordinary global mimetics in reporting and commenting upon the world sources of spiritual trouble.[14] The pedagogical challenge this involves quite exceeds anything we can reasonably expect in a postmodern age of collapsed social narratives. At the same time, emancipatory ideologies of the absolute autonomy of the individual in all matters of sexuality and reproduction cannot be pursued outside of a framework of institutions, law, and morality, which in turn require large allocations of public energy.

The crises of global culture are at once extraordinarily nation state-building, at least with respect to the levels of the therapeutic apparatus of the State—and transnational on the economic and political level, depending upon shifts in the multinational corporate agenda. Of course, these two levels interact, so that global capitalism responds to its own trans-state activities through the nation state and even through a layer of "international" agencies. Seen in this context, AIDS is again simultaneously a globalizing panic and a national state epidemic, mobilizing government health institutions from the municipal level all the way to the World Health Organization.[15] As a global panic, AIDS becomes a further charge upon the Third World, whereas US AIDS is principally an advanced economy, urban male (age 20-40 years) anal partner hazard, with drug users and bisexual males as secondary transmitters. US AIDS has benefited from the same trade routes as other sexually transmitted diseases. The imperial dominance of American capitalism within the global system, of course, diffuses American "life style" ideologies through global mass culture, tourism, commercial, and military travel, as we have seen in an earlier chapter.

Thus, the global health system is only the promissory side of a world disease system. Each generates the other. Here, once again, there is a potential for a rebarbarization of the global order through quarantine orders, immunization control, and racism—witness the construct of AFRO-AIDS. The concept of AFRO-AIDS is marvelously suited to project on to "world history" so to speak, an "Afro-origin" for AIDS whereas the socioeconomic conditions of Blacks and Hispanics in the United States and its dependencies are clearly the principal source of American disease, crime, and poverty. Similarly in Africa, where AIDS is a widespread heterosexual trouble, one must take into account shifts in marriage practices due to urban migration, poverty, and a fragile medical infrastructure, before "racializing" the disease.[16] Yet the search for a Simian-based HIV, endorsed by the *Scientific American*,[17] inspires hopes of naturalizing a colonial and class history whose overwhelming dimensions cannot possibly be reduced by the biomedical sciences. Since America's internal black population is the immediate source of virtually inorganizable and ineducable IV drug users, who are in turn viewed as the principal transmitters in the heterosexualization of AIDS, a third crisis looms within the US medical system inasmuch as so many millions of Americans are without any medical coverage, but could hardly continue to be ignored should their deaths be attributable to AIDS.

As we have said, AIDS as a global pandemic puts considerable stress on the international (world) health order. The United States and Africa are the two epicenters of the AIDS pandemic, with respectively 204 and 150 cases per million, despite huge differences in socioeconomic and sociomedical infrastructures. Yet these two centers cannot be treated in the same way, as though the one were the shadow of the other, nor can they be allowed to drift apart. In the meantime, the world medical order will have to reappraise its foundations built upon Western biomedicine and the colonial power of its corporate pharmacological institutions which have been shamed into reducing the price of treatments to Africa only in 2001.[18] National states will vary in their capacity to sustain the costs of AIDS where these are predicated upon a purely medical strategy that presupposes no state intervention with respect to high-risk sexual and narcotic behavior. These countries may also take different stances on American efforts to medicalize AIDS, just as they may or may not cooperate with American efforts to "police" the international order on such other issues as immigration, or environmental pollution. Similarly, it may not be possible to impose allegedly international standards of medical and social science research across cultures whose definition of illness, community, and knowledge are known to vary. To do so embroils such research in counter charges of a *medical imperialism* with which colonial countries are already familiar.[19]

It is a conceit of the American political order that beyond its borders life is everywhere short, nasty, and brutish—despite the fact that its own urban scene answers at least as well to such description as any foreign culture to which it is thought to apply. US AIDS intensifies the lethal content of American culture in unprecedented ways because it threatens to spill over the class wall that separates the rich from the poor, the suburbs from the inner city and family life from individual life-styles that challenge it.[20]

By the same token, since this spillage has largely been a construct of the media coverage of AIDS, the state therapeutic complex has simultaneously achieved a considerable "containment" of the epidemic as one that is by and large responsive to its administrative institutions without raising revolutionary changes to our fundamental ethical and political constitution. So long as we are able to muddle along in this fashion, we avoid the most catastrophic scenarios envisaged as an effect of the global ravage of AIDS. Here, of course, the ultimate breakdown would be in *the class system as an immunological order* and the destruction of the medical system predicated upon such an order. Short of such a conflagration, we may expect class politics to slide into caste politics in the hope of preserving the health of society by sacrificing its principle of charity to group preservation. To the extent that this scenario develops, the global order will have collapsed into a barbarous conflict of national biological elites, each seeking to preserve its own purity while trying to eliminate the other as a possible contaminant.

NOTES

1. Paula A. Treichler, "AIDS, Homophobia, and Biomedical Discourse: An Epidemic of Signification," in *AIDS: Cultural Analysis Cultural Criticism*, Douglas Crimp, ed. (Cambridge: The MIT Press, 1988), p. 68.

2. *AIDS: A Perspective for Canadians* (Ottawa: The Royal Society of Canada, 1988); David Spurgeon, *Understanding AIDS: A Canadian Strategy* (Toronto: Key Porter Books Limited, 1988); *Confronting AIDS: Directions For Public Health, Health Care, and Research* (Washington, DC: National Academy Press, 1986); *AIDS: The Burdens of History*, Elizabeth Fee and Daniel Fox, eds. (Berkeley: University of California Press, 1988); Dennis Altman, *AIDS in the Mind of America* (New York: Anchor Press, 1987); *"Living with AIDS,"* Special Issue of *Daedalus* (Spring 1989). Cambridge, MA: American Academy of Arts and Sciences; *New England Journal of Public Policy*, 4:1 (Winter/Spring 1988): Special Issue on AIDS; *The Milbank Quarterly*, 64 (1986): Special Issue on AIDS: The Public Context of An Epidemic; Michael Ornstein, *AIDS in Canada. Knowledge, Behaviour and Attitudes of Adults* (Toronto: University of Toronto Press, 1989).

3. "What Science Knows About AIDS," *Scientific American*, 259:4 (October 1988); see also *Living with AIDS*, op. cit.

4. *AIDS: Cultural Analysis Cultural Activism*, Douglas Crim, ed. (Cambridge, MA: The MIT Press, 1988); Simon Watney, *Policing Desire: Pornography, AIDS and the Media* (Minneapolis: University of Minnesota Press, 1987).

5.   John O'Neill, *Five Bodies: The Human Shape of Modern Society* (Ithaca: Cornell University Press, 1985).

6.   Gayle Rubin, "Thinking Sex: Notes for a Radical Theory of Politics of Sexuality," in *Pleasure and Danger: Exploring Female Sexuality*, Carole S. Vance, ed. (London: Routledge and Kegan Paul, 1984), pp. 267-329; Robert Crawford, "You are Dangerous To Your Health: The Ideology and Politics of Victim Blaming," *International Journal of Health Services*, 7:4 (1977):663-680.

7.   *The Global Political Economy: Perspectives, Problems and Policies*, David Law and Stephen Gill, eds. (New York: Harvester Press, 1988).

8.   See Chapter 8.

9.   John O'Neill, "Religion and Postmodernism: The Durkheimian Bond in Bell and Jameson," in *Theory and Culture and Society*, Special issue on Postmodernism, 5:2-3 (June 1988):493-508.

10. See *Five Bodies*, Chapter Four, "Consumer Bodies."

11. Allen M. Brandt, "AIDS: From Social History to Social Policy," in *AIDS: the Burdens of History*; and Dennis Altman, "Legitimation through Disaster: AIDS and The Gay Movement," also in *AIDS: The Burdens of History*.

12. John O'Neill, "The Medicalization of Social Control," *The Canadian Review of Sociology and Anthropology*, 23:3 (August 1986):350-364.

13. Paul A. Treichler, "AIDS, Gender, and Biomedical Discourse: Current Contests for Meanings," in *AIDS: the Burdens of History*, and her "Biomedical Discourse: An Epidemic of Signification," in *AIDS: Cultural Analysis Cultural Activism*.

14. Timothy W. Luke, *Screens of Power: Ideology, Domination, and Resistance in Informational Society* (Urbana: University of Illinois Press, 1989), Chap. 3, "From Fundamentalism to Televangelism."

15. Nicholas A. Christakis, "Responding to a Pandemic: International Interests in AIDS Control," in *Living with AIDS; and Confronting AIDS*, Chap. 7, "International Aspects of AIDS and HIV Infection."

16. Alfred J. Fortin, "The Politics of AIDS in Kenya," *Third World Quarterly*, 9:3 (July 1987):907-919; Nicholas A. Christakis, "The Ethical Design of an AIDS Vaccine Trial in Africa," *Hastings Center Report*, 18:3 (June/July 1988):31-37.

17. Max Essex and Phyllis J. Kanki, "The Origins of the AIDS Virus," *Scientific American* (October 1988):64-71.

18. Charles Hunt, "AIDS and Capitalist Medicine," *Monthly Reviews* (January 1988):11-25; Vicente Navarro, *Medicine Under Capitalism* (New York: Prodist, 1976); B. Ehrenreich and J. Ehrenreich, *The American Health Empire: Power Projects and Politics* (New York: Random House, 1971).

19. See Christakis, note 13 above.

20. John O'Neill, "Two Cartographies of AIDS: The (In)describable Pain of HIV/AIDS," in *When Pain Strikes*, edited by Bill Burns, Cathy Busby and Kim Sawchuk (Minneapolis: University of Minnesota Press, 1998), pp. 292-308.

CHAPTER TEN

Televideo ergo sum:
Some Hypotheses on the Specular
Functions of the Media

In addition to the arguments advanced in Chapter Five on communication and the legitimation problem, I propose to extend that discussion with the aid of some propositions drawn from critical theory, psychoanalysis, and semiotics, to show how we are to interpret the political, economic, and social integration of the media. I shall argue that the media simultaneously dehistoricize and defamilize production and consumption for the masses, while they refloat the professional and political legitimation of these processes. Thus *the prime specular function of the media is to close the ideological circuit between the visible and the invisible society.*

THE DREAM OF THINGS

The lord of creation is drowning in his own act. We are swamped by things. We own much but find that we do not possess ourselves. A curious solution to the problem of scarcity—abundance as freedom nevertheless fails us. Go back to the beginning. Man is the measure of all things. In the days when the body was simple, things were simple. The body as a state of nature the naturally scenic truth of things not to exceed its needs. Somehow the body lost its senses. It became excessive. Wants took off from needs. The body became a pandora's box with global dimensions, mythic at first, and by the same token or coinage, turned television—the box that is always on. Eye of the world, tireless machinery of the visible and the invisible.

Is there no exit from the dream that television dreams for us? Our answers are likely to be more ready the less we explore the functions of television's dreamwork. Why should we sever ourselves from the warm and generous light of television? After all, television is at our service. It disgorges nothing but what we want and it does so without the seasonal moodiness of mother earth. Its corporate dispensaries are more paternal than our own fathers; its concerns on our behalf are endless. If it thereby succeeds in infantilizing us, hiding from us the realities of its political economy, it does so only to serve us. If it deceives us with what we see, it does so only to delight us. There is nothing malign or evil in this. No conspiracy.

If this seems fantastic, it is because we are fantastic, not because television manufactures our dreams without us. To believe that television invades our minds is to oppose the old rational myth of the sovereign consumer in a world of objects enumerable like the animals on Noah's Ark. It is to forget that the media are the extensions of our bodies, the means of our multiplication, whereby the body escapes the prison of the mind and desire conquers reality. If we are not liberated in the expansion of desire, any more than we are made knowledgeable by the expansion of information, it is because the essential activity of televised passivity is consumption, the hypnotic possession of powers that freed themselves only in imagination. If we lose our public will in all this, it is because television universalizes our most intimate desires, collecting us in the kingdom of cleanliness and self-rewarding justice, entered only by those who are television's children:

> The art of consuming is as much an act of the imagination (fictitious) as a real act ('reality' itself being divided into compulsions and adaptations), and therefore metaphorical (joy in every mouthful, in every perusal of the object) and metonymical (all of consumption and all the joy of consuming in every object and every action). This in itself would not matter if consumption were not accepted as something reliable, sound, and devoid of deception and mirage, but there are no natural frontiers separating imaginary consumption or the consumption of make-believe (the subject of publicity) and real consumption; or one might say that there exists a fluid frontier that is always being overstepped and that can only be fixed in theory. Consumer-goods are not only glorified by sign and "good" in so far as they are signified; consumption is primarily related to these signs and not to the goods themselves.[1]

Television is our body, our vision, our mind, our sanity, our appetite, our will to live as we do. Television is our mother's body, endlessly materializing our appetite for security, nostalgia, and happiness. Television is our father's body, in heaven and on earth, patient and strong, loyal and free. Television is our brother's body, subversive and superfluous, the joker in the pack. Television is old bodies, black bodies, broken bodies, bodies resurrected, imprisoned, and burned. Television is Christ's body sacrificed in the same way every day everywhere in the world with the same benign indifference to the local sufferings, ignorance, injuries, and fears that it takes upon itself *per omnia saecula saeculorum*. It is to this god that we offer our detergents, our deodorants, our dog food, toothpastes, and beer that He take upon Himself our murders, rapes, deceptions, and insanity—*agnus dei televisionis*. Thus our metabolism turns symbolism in an unending Mass celebrating the bonds of everyday life before millions of families gathered at the altar of television. To realize this, things have ceased to be objects of use; rather, they exceed their uses in order to float as symbols of the desire-to-desire whereby ordinary men divinize themselves; thus the modern soul drifts on things, trusting to them, like a man sold to his own hi-fidelity system.

Things no longer serve us. We are all the sorcerer's apprentice. Knowledge no longer serves us. Class, like the encyclopedia, is likewise an outworn container burst by the proliferation of things as signs, images of ourselves that no longer center upon ourselves. Things continue to mutate, expand, self-destruct; there is nothing outside of this wild economy of signs flashing between signs. Of course, there are catalogues, texts that strain to reclassify things in terms of size, price, function, seeking thereby to tie them back to the referentials of the everyday world. Things, after all, must be generated according to the natural places of use-value—a demand that science no less than commonsense continues to make upon them. Yet phenomenologically we remain as ignorant of the technical constituents of things as we do of their nutritional contents. Let the machines talk to themselves of their inner lives, so long as they leave us their digital signs to play with. We are the absent landlords of technology, eating our way through the cornflakes to win more gadgets or to become a body burning in the distant sun of fly-away happiness.

Thus our categorical schemes do not hold; there is no center to the modern political economy of things and their way of grafting upon everyday life. The engineer and the consumer sovereign are distant ends of an economy whose uses run to the irrealization of the real in the necessity of the unnecessary. In a carnival of the impossible, the engineer and the consumer divide between them the sense and nonsense of political economy. Each continues to think of his activity in terms of function and need, while recognizing that these are held hostage in an extravagant world that outruns its internal poverty and even forsakes its own sanity. The *pseudodifferentiation* of things is entirely dependent upon television and advertising to transform it into *pseudochoices*. *Pseudodifferentiation* materializes abundance and choice. It thereby makes the consumer necessary to its act, seeks to advise and persuade its sovereign to make rational decisions in the very world it has made irrational.[2] Moreover, the same split is achieved as an ecological setting. The things that surround us are catalogued, shelved, inventoried, enumerated, and every effort made to restore them to their place if lost or stolen. Things are timed, weighed, and priced to ring the bell of modern accountability. Such things are adored, watched reverently in the windows of the economy, and even visited by tourists in the birthplace of their production. Such things celebrate themselves, animating our lives and our homes as a continuous avalanche of novelty, happiness, and utility.

Television fulfills itself in the show and tell of political economy.[3] *The commodity fetish is a talking, seeing, feeling god inviting us into the liturgy of consumption to dispel the sorrows of production.* Nobody wants to work in an ice-cream factory. But everyone loves the ice cream shop as the dispensary of frivolous choice and pleasure, the natural setting of industrial freedom and the Americanization of happiness. Therefore things have their own life, their colors, shapes, surfaces, materials, and ambiance. And they do so because their life is our life. Because this is so, things necessarily sing themselves to us in the daily celebration of the life we have in common. The loveliest of these things used to be the automobile, our home away from home, suspended, floating, new, shiny, sexy, and safe. This work-horse sings of itself as a centaur, a space invader, conquering everything from the forest to the beach. Everyone wants to float in an automobile, but no one wants to see the sorrows of the assembly line or the traffic jam. Thus if public transport is eclipsed by the automobile, it is not because it is less efficient, but precisely because it is merely an efficient mode of transportation, a concession to the body that is hard on dreams. By contrast, the automobile is also the perfect expression of the body that dreams, the narcissistic and phallicized will-to-power that bends to no one but the pedestrian and the policeman who oppose to it the primal body in a civilizing risk of murder. Today, the computer is the ultimate escape machine.

Commodities, like men, praise themselves. In either case, the praise is most loud when men and things are dead. Once our ancestors no longer compete with us, we remember their presence. In a world of garbage, such worship would seem to escape things; or it would, were it not the case that things are able to intercede in the world of men. Thus we collect things we call antiques or originals, as memorials of their origins and collective life which they exempt from use and decay, setting them apart in their homes and museums like church relics. The town collects the countryside, and industrialized societies collect nature societies, arresting their history if necessary, in order to preserve a mirror of the civilized past. Such collections are produced by force and murder, to be enjoyed in peace and happiness at home or in decadent displays of tourism abroad.

The phenomenology of everyday life is largely a spectacle of things. However, this spectacle is not opened up to any transcendental viewer. In other words, we do not watch television in order to see things but rather we watch things in order to watch television or the Net. This is the heart of our passivity. For the more we watch ourselves, the less we see of ourselves. That is the price of our free admission to the theater of things. By the same token, we are obliged to take our own place in the parade of things. Here we work out a marvelous symbiosis between necessity and choice. Things personalize themselves; they separate by price, quality, color, and ambiance in order to allow us to insert our most intimate self into them while at the same time assuring us of a common fate. In this way, things enable us to make a choice on their behalf that has already been made for us. If there is any inconsistency in this, it is the necessary sacrifice of logic to myth, since we wish to honor the twin gods of consumption and production. To do so, we sacrifice the sovereign consumer and the rational engineer to the self-improvement of things that improve ourselves.

In the fashion show and the auto show man's two bodies are celebrated today for their embrace of the appearances that will degrade them tomorrow. But the god of fashion is never so honored as by those who risk being out-of-fashion by being-in-fashion. For the existential risk, whatever Sartre may have thought, lies between Being and Appearance, not between Being and Nothingness. It is this difference that we monitor in ourselves through our television: *televideo ergo sum.*

Thus before/for us daily consumption assumes its dual aspect and its basic ambiguity. Taken as a whole, quotidian and non-quotidian, it is material (sensorial, something to be taken, used, consumed, experienced) and theoretical (or ideological-images, symbols, signifiers, language and meta-language being consumed); it is complete (tending towards a system of consumption based on the rationalized organization of everyday life) and incomplete (the system is forever unfinished, disproved, threatened, unclosed, and opening on to nothingness); it is satisfaction (of needs, this one or that one, the need for this or for that, therefore sooner or later it is saturation) and frustration (only air was consumed, so the desire reemerges); it is constructive (choice of objects, ordering, filing, contrived freedom) and destructive (it vanishes in the center of things, slides down the slopes of piled up objects accumulated without love and for no purpose). The so-called society of consumption is both a society of affluence and a society of want, of squandering and of asceticism (of intellectuality, exactitude, coldness). The ambiguities proliferate, each term reflecting its opposite (its precise opposite, its contradiction, its mirror-image); signifying it and being signified by it, they stand surety for and substitute each other while each one reflects all the others. It is a pseudo-system, a system of non-systems, cohesion of incoherence. The breaking point may be approached but never quite attained: that is the limit.[4]

Today we are born and die on television and computer screens. We mourn, starve, rave, and pray on television. Even the Pope has discovered that the world's roads need no longer lead to Rome so long as he can make the news. Thus television reaches ever greater heights of efficacy. For now theologians need to muse whether the Papal Blessing reaches those who watch it as well as those who kneel in His presence. And even Freud might wonder whether live assassinations reconfirm his fantasy of the historical murder upon which all civilization is founded. When Presidents, Queens, and Popes run this risk before world television, who is to say where reality and appearance begin or end? Perhaps the only test is whether these events are punctuated with commercials. If television is bad for anyone, it seems to go worst for Marx, unable to defend himself against even Frank Sinatra. If the leaders of old Europe could be dragged into Reagan's chorus line for Let Poland Be Poland, one can argue that was to defend their own domestic use of propaganda rather than from any faith in television as *realpolitik*. Yet viewed from another angle, this *festival of public opinion* represents an extension of air power that has always fascinated politicians in their war for minds. Thus the Presidency of Ronald Reagan was not embarrassing because he was an actor. It was embarrassing because his executives had still not yet discovered how to make politics show business. Their successes and failures in this direction are, however, tied to the same processes at work in the economy. Today, the economy has internalized the media and by the same process is subject to media (re)presentation.

THE SPECULAR FUNCTION OF THINGS

To proceed with my argument I need, first of all, to say something about the notion of *specularization* as I employ it, and then to connect it with the *theory of ideology* and *commodity fetishism*. What makes television the connecting link here is understandable only if we first situate media technology within the semiotization of all modern technology.[5] We can no longer think of technology as a neutral instrument of mediation between man and nature. Modern technology is increasingly assimilated to the reciprocal adaptations of man, nature, and society which occur in a horizontal field of information or communication. Within this field, objects, values, and experiences of every kind function as interchangeable elements of a semantic field in which reality, persons, events, and things are reduced to the possession of style.

What is now needed in media studies is a *structural theory of fetishism*—analogous to the psychoanalytic theory of the structure of perversion—functioning on the same level as Marx's structural theory of the commodity process.[6] This involves recovering the original sense of the practice of making fetishes. Originally the fetish is an artifact made to imitate something else through the signs and marks inscribed upon it. The fascination is with the power of the signs to manipulate forces beyond the fetish. Thus what we mean by *commodity-fetishism* is not any desire for objects or experiences as such but the *desire of desire*, that is, desire of the structural codes through which desire of any sort is mediated. Commercial desire is not stimulated by the novelty of objects and experiences, or by the need to fill expanding leisure time. Desire derives from the continuous expansion of the system of commodity-signs towards which the consumer is directed by means of the media which furnish up-to-date readings, or *messages assembled within-and-as the settings of everyday life*. It is here that *models*, *styles*, and *collections* (see any large store's catalogue) function as structures of desire, as absolute moments in a stream of commodities.

Moreover, commodity fetishism must deal with the *body politic*, that is to say, with family health, working life, and personal happiness, since these must be refloated as the basic referentials/representations of the social structure (class, race, gender, ethnicity) through which the specularization of the commodity system functions. We can then realize how the media are obliged to beautify and transfigure the body and its everyday settings in order to *re-write the gap between reality and desire* within a self-sufficient system of symbols invulnerable to individual failure or class limits.[7] It is this self-reflecting specularity of the commodity system that constitutes its fetishistic power to engage us in its charm, fun, and fantasy.

Television is perfectly suited to the specular work of breaking down the tension between public and private life, work and consumption, vicariousness and authenticity. All public and personal relations are now mediated through the communicative pattern of objects and experiences transfigured in the lights, sounds, colors, and scenarios of media technology. By projecting the passive viewer into the spectacle of his or her own everyday life as an environment from which all signs of the structure of social and political domination are *erased*, television hides the system.

The specular function of things—their symbolic proliferation, unbounded by space and time, their instant and imaginary availability— is *to make modern political economy impossible to understand.* It is to hide society as a circus or carnival of things. Above all, it ties things to the life of consumers, subordinating the world of production to the service of desire.[8] If there is a real world that sets any parameters upon the economy of desire, it arises only in the contingencies of domestic and international violence. Hence television regularly interrupts its celebration of things with the news of things that are recalcitrant to commercial celebration—such as war, murder, and the weather. In such portrayals of anarchy and scarcity, it reconfirms the homogeneity of its own comfortable world. It also achieves a *pseudocritical stance*, modifying its own integration of desire and the unconscious imagination which ordinarily functions to pacify criticism. Thoughtfulness is absorbed into the documentary and dialogue is enshrined in the talking portraits of the ABC News. The result is a perfect combination of docility and aggression—familiar staples of television. This combination is so potent that only Walter Cronkite and Barbara Walters could give it a family face. Thus television, like the Kremlin, has its succession problems: Nothing must destabilize the continuity of the image of stability in which America is anchored. Indeed, this formula appears so potent that Canada had to transform its national news center into a spaceship transmitting world events through the gentle body of Knowlton Nash who in turn handed over the infighting to his avenging daughters Barbara Frum and Mary Lou Finlay. Since Knowlton was to Barbara as Barbara was to Mary, one could predict that Mary would disappear from The Journal, as it was called, in order to restore the heterosexual balance of truth and history in our nation. In fact, this came to pass.

World News reporting, if anything, might seem to confirm for us McLuhan's thesis that through the medium of television we all inhabit a global village. In my view, McLuhan's concept of the global village merely grasps superficial features of the new body politic without being able to relate them to any analytic framework of the ties between the modern administrative state and the political economy whose specular processes are serviced by the media. It is, nevertheless, hazardous to try to generalize upon the basic political functions of the media. The media employ a variety of audiovisual technologies, ranging from satellite television to newspapers, magazines, scientific journals, catalogues and handbills, each of which can play a specific role in raising or depressing the level of public awareness. Thus the specular functions of the media are at first sight extremely varied and not easily brought to order. Yet, as I have argued in Chapter Five, it may be useful to think of the media as functioning to *specularize the legitimation processes of state administered capitalism.* The latter necessitate the use of television in the production of the following spectacles:

1. On the *state level* the media communicate state or national issues, the "national" news, and the national parliamentary or congressional proceedings;
2. On the *market level* the media continue to portray the market system as though it were the basic legitimating agency of bourgeois values, and encourage the pursuit of consumerism and leisure;
3. The function of the media at both the state level and the market level is to manage the necessary separation between expert and technical administration processes, on the one hand, and public participation, on the other;
4. This is achieved through the media-fix of turning the political and economic processes into spectacles, while at the same time raising the level of *civil privatism* and *depoliticizing the public realm*;[9]
5. To the extent that the media can be used to manage the legitimation problem at the market level, i.e., as *sociological propaganda*, then the state use of the media for political mobilization is reduced, and appears as a democratic instrument of information: *the right to know*;
6. In turn, the media generally assume an apparently benign function in filling the leisure time of a highly privatized mass democracy. By and large, the communicative process of *popular education* consists of the mobilization of persons into a structure of self-coercion and domination exercised against nature (*education for work*), against human nature (*education for living*), and against the collective potential for freedom and expression (*political education*).

It is important to see that the specular function of television lies in its capacity to individualize the mass while treating the individual only as a member of the masses. It achieves this as a machinery of *sociological propaganda* which works upon individuals from the bottom up, so to speak, rather than through the imposition of ideologies to influence their minds:

Such propaganda is essentially diffuse. It is rarely conveyed by catchword or expressed intentions. Instead it is based on a general climate, an atmosphere that influences people imperceptibly without having the appearance of propaganda; it gets to man through his customs, through his most unconscious habits. It creates new habits in him; it is a sort of persuasion from within. As a result, man adopts new criteria of judgment and choice, adopts them spontaneously, as if he had chosen them himself. But all their criteria are in conformity with the environment and are essentially of a collective nature. Sociological propaganda produces a progressive adaptation to a certain order of things, a certain concept of human relations, which unconsciously molds individuals and makes them conform to society.[10]

Sociological propaganda reaches minds through bodies; it works especially upon familied bodies and the tissue of everyday life. Here the specular function of the family soaps may be understood. *Soap operas are essentially defamilizing dramas*, whatever the folklore on them. They are essential to the work of television, occupying as much viewing time as the working week. In soap operas the family is on stage because it is offstage breaking up, falling apart, distrustful, and anomic. Children are largely absent from family soaps because the family is no longer sure of its commitment to reproduce itself. Soap wives either have trouble conceiving by their husbands or else the embryo has even more trouble getting born, as if aware that family life may no longer be the blessing it used to be. Surviving infants turn to a life of illness, accidents, and tragedy, a concern to their helpless parents, whose lives provide no strength because they are always crippled with divorce and infidelity. If the child learns to cope with parents, it learns by coping with their serial marriages. The child's family life is a soap opera. If the soaps remain fun, it is because the props remain reassuringly the same, inhabited by other people's troubles. However, once these troubles turn into our own—as we half suspect they will—the soaps leave us unprepared after all and ready for the professional therapeutic care which is the last resort of the family.

Of course, television resists the bureaucratized therapeutic professions. It is democratic, maverick, and lawless. It favors people without titles, private individuals, common folk, comic children, and ethnics. Thus television is especially concerned to portray the private cop or detective who goes for his man with more passion and concern for speedy justice, more wit and strength than the fat, bureaucratic legal and criminal system. What I have in mind here may be illustrated from certain features of the series *Kojak*. Here we have the law and its pursuit of justice embodied in a fat, bald headed, lollypop sucking Greek. How is this possible? Anyone familiar with the rituals of purity and danger would realize that the symbolism of the law has been seriously polluted by its embodiment in a figure like Kojak, not to mention his equally obese aides and black colleagues. Or are we dealing with a liberal system of justice unfortunately obliged to prey upon an otherwise criminogenic ethnic community? To be brief, the answer is that Anycop's work has been simplified by the technology of crime search, namely, by the bugging devices that permit criminals to tell the police where to look for them. This device raises the efficiency of the search and relieves its parties for the symbolic work of token ethnic representation. As a matter of fact, it even permits policemen to turn into policewomen! My point is that the technology of modern society is indifferent to gender, race, and creed, but may be used to "refloat" them as though they were now the local, human, and vital elements of modern life. "Multiculturalism," which is the Canadian version of this phenomenon, merely floats ethnicity in an *imaginary festival* of consensus. It is an engaging symbolism of liberal consensus, an aesthetic resolution of the endemic conflicts in what remains a stratifying society.

Between them the criminal and the entrepreneurial cop revive the American free enterprise system in settings that celebrate its technology, affluence, and ultimate goodness. By the same token, the lid of authority and professional, bureaucratic power can always be replaced in dealing with the ultimate threats to family and corporate life. Thus crime, illness, disaster, and intrigue are sacrificed on the altar of television in order to reconfirm the justice and goodness of the institutions that defend the family against violence and evil. But *if television defends the family, it does so only in order to redesign it.* Thus in the food commercial, television makes us hungry, teaches us what to eat and how to eat. Television even admonishes us for overeating what it offers us to eat. This is because television is in the business of defamilizing the production and consumption of food. To achieve this, it is necessary to decathect housework, cooking, and cleaning and to install the processed, precooked, and self-cleaning substitutes from television's fantasy world. Food must be reduced to energy; living, too, must be reduced to an energy calculus, fast and efficient. Anything short of this is idiocy and old age. Once the production and consumption of televised food has been defamilized, the symbolism of home, mother, tender and loving care can be refloated as literal ingredients of the substitute foods. It is then ready to be eaten by children without parents, husbands without wives, and working mothers without time. The side effects can be taken care of in the stomach aid and breath conquering advertisements. TV dinners are the totem meal of televisioned living.

The specularization of consumption, coupled with the erasure of production, is marvelously suited to the *materialization of the ideology of equality and happiness.* Television democratizes consumption and is silent upon the stratification of power and privilege generated and reproduced in the system of production. Indeed, the stratifying processes in the production system are resymbolized as levels of taste, desire, and adventure open to advertising. In a world where access to health, education, employment, housing, and food is manifestly unequal, television pictures consumption as a magical school in which we learn to want things for their fetishistic power to alter and improve our lives. Television thereby erases the fact that consumption is part of the system of scarcity/affluence rather than the delivery of abundance. *Television schools us in envy* without the desire to learn anything about the system of deprivation that floats envy, comparison and competition with ourselves, or competition with others. It is for this reason that television feminizes consumption, selling to women the myth of woman, just as it masculinizes consumption, selling to men the myth of man. In this way, human relations are completely relativized, since the logic of capitalism is indifferent to all relationships except as they can be refloated in the system of exchange value.

The specular function of television is therefore to conceal the differentials, comparisons, and exclusions that separate men by surrounding them with the same swarm of images. In other words, it is to broadcast an ideology in which individuals are at the center of things and images, while being absent from the processes that shape these images of their everyday life:

> We can define *the field of consumption*: it is everyday life. The latter is not just the sum of daily events and deeds, the domain of banality and repetition, it is a *system of interpretation*. The everyday consists of the dissociation of a total praxis into a transcendental, autonomous, and abstract and immanent "private" sphere. Work, leisure, family, friendship are all recognized by the individual in a static mode, outside of the world and history, as a coherent system grounded as the closure of the private, the abstract liberty of the individual. . . . From the objective standpoint of the totality, the everyday life is an impoverished and residual sphere. But it is also triumphant and euphoric in its efforts to achieve a total autonomy and a reinterpretation of the world 'as an internal usage.' Herein lies the profound and organic symbiosis between the privatized sphere of everyday life and mass communications.[11]

Mass media culture is *fictional* whether it is delivering information or art, since what it is shaping is not just its images but the kind of public that is organized, that is, privatized through the media. The media do not present the public with information with which to determine their political and economic choices. Rather, the media relay the choices already made in the political economy for *pseudoratification* by the viewer-consumer-voter. Thus we can understand, for example, the symbiosis between the self-expanding field of complex information systems and the talk or gossip show. How does the latter thrive in such a world? First of all, the talk show specularizes a dying art. Moreover, it only appears to be an intimate oasis removed from the information system. This is because its gossip revolves around the same values of power, money, and sex as rule the conglomerate world and so convey the same basic message on behalf of the system. The talk show, then, is the intimate end of the machine that never ceases to talk about itself, that is always on because *television is the sun that never sets on the empire of commerce.* Indeed, the whole point is that television is always talking because it never talks *to* us but *at* us. It is a necessary communication in this form since it reproduces the unilateral power structure in which it is embedded. We are all joined in television because we are all separated from knowledge and power in the specularization of information, desire, and intimacy. *Television circulates our isolation.* It does so by embracing us in an imaginary community without local and national boundaries. Television is everybody's sexual partner and the nation's confidante. Your personal computer will confirm this to the very end of your finger tips.

CONCLUSION: SOME HYPOTHESES ON THE SPECULAR FUNCTIONS OF THE MEDIA

McLuhan notwithstanding, the media do have a message and it is the message that makes the media necessary as instruments of global domination. The basic functions of the media messages are:

1. To float the real worlds of production and consumption in a semiotic system of signs without referents;
2. Within (1) to homologize psychic and social structures;
3. By floating an autonomous semiotic field in life-styles, models, collections, news, information, to hide the system of administrative controls operating on (1-2).
4. To reduce everyday life to a privatized celebration of the benign products of capitalist democracy;
5. To dramatize in individualized morality plays (including psychoanalysis) which save the system the failures, tensions and reality gaps in the system of administrative controls;
6. To produce political and social amnesia;
7. Not to charge consumers for (1-6). That is why the message, if not the medium, is free. Where else—or why else—can one get something for nothing?

The specular tasks of the media are necessary functions in the legitimation of capitalist democracy. They enable the system to employ sociological propaganda, thereby minimizing more overt political mobilization. To achieve this, the following spectacles are necessary media functions:

1. a. to communicate state or national issues;
    b. to communicate the national "news";
    c. to televise the national parliamentary or congressional proceedings;
2. a. to continue to portray the market system as though it were the basic legitimating agency of bourgeois values;
    b. to encourage the pursuit of consumerism and leisure.
3. a. expert and *technical administration* processes, and
    b. *public participation*
4. The requirements of (III) are serviced through the media-mix of
    a. *spectacularizing* the political and economic process, and, at the same time,
    b. raising the level of *civic privatism* and *depoliticizing the public realm*.
5. To the extent that the media can be used to manage the legitimation problem at level (II), that is to say, as a market communication, then the state use of the media assumes the form of *sociological propaganda*. In turn, the media generally assume an apparently benign function in fulfilling the leisure time of a highly privatized mass democracy.
6. By and large, what we call mass culture and democracy consists of the mobilization of persons into a structure of self-coercion and domination exercised
    a. against *nature*, as the raw material of industry;
    b. against *human nature* itself, as a primary source of pleasure and gratification;
    c. against the *collective potential for freedom and expression* historically realizable on the basis of the structures of organization and repression in (a) and (b).
7. In short, industrial democracies, as we have known them so far, are the product of a *triple structure of domination over nature, persons, and society*, realized through the instruments of technology, repressive socialization, and the class expropriation of collective political power. This is the matrix of media work.
8. Media work is nevertheless not engaged in a great conspiracy. It gets its job done because in a certain sense *the media are always the media we deserve*. Television puts individuals in communion with themselves while leaving them at their relatively isolated terminals. Together, then, the message and the medium, the audience and the communications industry are united around TV as their culture's *bard*[12] whose daily task is to compose a series of messages that communicate to the audience a self-confirming vision of themselves and their everyday beliefs and values.

To do so, TV must relay the authority of the community's myths rather than the genius of its producers, an effect it achieves by affirming orality and presence against rationality and absence, or by defending primary institutions in their struggle with secondary institutions.

9. The *bardic functions* of television may be summarized as follows:

   a. To *articulate* the main lines of the established cultural consensus about the nature of reality (and therefore the reality of nature).

   b. To *implicate* the individual members of the culture into its dominant value-systems, by exchanging a status-enhancing message for the endorsement of that message's underlying ideology (as articulated in its mythology).

   c. To *celebrate*, explain, interpret, and justify the doings of the culture's individual representatives in the world out-there; using the mythology of individuality to claw back such individuals from any mere eccentricity to a position of socio-centrality.

   d. To *assure* the culture at large of its practical adequacy in the world by affirming and confirming its ideologies/mythologies in active engagement with the practical and potentially unpredictable world.

   e. To *expose*, conversely, any practical inadequacies in the culture's sense of itself which might result from changed conditions in the world out-there, or from pressure within the culture for a reorientation in favour of a new ideological stance.

   f. To *convince* the audience that their status and identity as individuals is guaranteed by the culture as a whole.

   g. To *transmit* by these means a sense of cultural membership (security and involvement).[13]

10. TV HEALTH WARNING

    a. There is never anything on TV—if there were, the system would break down;

    b. Since all of life is on TV, you can expect to have already seen it;

    c. TV is part of the family—it babysits its mind;

    d. TV is not free—you must give your time to it;

    e. No one knows whether or not TV is real.

NOTES

1. Henri Lefebvre, *Everyday Life in the Modern World*. Trans. by Sacha Rabinovitch (London: Allen Lane, 1971), pp. 90-91.
2. Jeffrey Schrank, *Snap, Crackle and Popular Taste: The Illusion of Free Choice in America* (New York: Delacorte Press, 1977).
3   Rose K. Goldsen, *The Show and Tell Machine: How Television Works and Works You Over* (New York: The Dial Press, 1977).
4. *Everyday Life in the Modern World*, p. 142.
5. Jeremy J. Shapiro, "One-Dimensionality: The Universal Semiotic of Technological Experience," in *Critical Interruptions: New Left Perspectives on Herbert Marcuse,* Paul Breine, ed. (New York: Herder and Herder, 1970), pp. 136-186.
6. Jean Baudrillard, *Pour une critique de l 'economie politique de signe* (Paris: Gallimard, 1972); Guy Debord, *La societé du spectacle* (Paris Buchet/Chastel, 1967); Robert Goldman, *Reading Acts Socially* (London: Routledge, 1992).
7. John O'Neill, "The Productive Body: An Essay on the Work of Consumption," *Queen's Quarterly*, 85(2) 1978:221-30.
8. See Stanley Aronvevitz, "Colonized Leisure, Trivialized Work," in his *False Promises: The Shaping of American Working Class Consciousness* (New York: McGraw-Hill Book Company, 1973).
9. Jürgen Habermas, *Legitimation Crisis*, trans. Thomas McCarthy (Boston: Beacon Press, 1975).
10. Jacques Ellul, *Propaganda: The Formation of Men's Attitudes*, trans. Konrad Kellen and Jean Lerner (New York: Vintage Books, 1973).
11 . Jean Baudrillard, *La societé de consommation* (Paris- Gallimard, 1970), p. 33; Gary Genosko, "The Paradoxical Effects of MacLuhanisme?" *Economy and Society*, 23, 1994: 407-432.
12. John Fiske and John Hartley, *Reading Television* (London: Methuen, 1978).
13. *Ibid.*, p. 88.

CONCLUSION

Reflections on the Day After Television

It is quite unthinkable for "the set" (video, computer) to go off. *Television is now an essential service.* No one could live without its 24-hour repair which underwrites our expectation that TV will always be there. The *repair myth* is essential to the technological myth that we shall never be abandoned by our machines. Because we can only imagine some evil genius cruel enough to destroy our machines, television is a prime machine for colonizing our imagination with scenarios of rebarbarization and intergalactic warfare designed to capture or to overwhelm our machines. The technological empire merely strikes back at its own nightmares.

Television, as Joyce Nelson has argued, is the perfect machine for the production of "atomic fictions"[1] in which we imagine our own destruction and post-TV resurrection. In this process we learn to accept "low levels" of radiation (higher for servicemen and nuclear towns) in kitchen and medical technology, and in television itself, as part of the quality of a life that is ultimately hooked into a nuclear weapons system within which we are all already dead. It is a curiosity of our civilization that we all insist upon using the best potatoes for french fries and fixing our teeth on the way to the nuclear oven. Thus TV has been the perfect *cold war machine* for portraying the American Way of Life as the staple of a world order which only a Third World War could disrupt. At the same time, this possibility is rehearsed in endless restorations of order on the domestic scene filtered through TV crime series and sitcoms in which America polices the world as it does itself on behalf of a blithe and befuddled American citizenry, docile at home and innocent abroad.

All Americans are held hostage to the American view of the world which very largely appears to them only through the American screen and its transmission of *RamboNews*:

> By controlling and dominating the screens of the world, the United States insists quite literally that its "vision" is the only one that shall prevail. Perhaps even more powerfully than its military supremacy, this techno-imperialism over the air-waves and the imagination of the world's population has caused, and still causes, extraordinary rifts and tensions as nations struggle against the overwhelming embrace of the global American agenda.
>
> . . . should we be surprised that so-called "terrorist" acts are specifically staged with television in mind? Since the American TV-screen shows virtually nothing from the rest of the world, while simultaneously deluging other countries with its own product, that U.S. TV-screen itself becomes the target for conveying, through even momentary news coverage, situations and conditions other wise excluded from the American frame of reference.[2]

Of course, it is quite easy to point to "serious" docudramas, panels and investigative journalism on TV which appear to challenge these generalizations. But the air of rationality affected by them is so thoroughly polluted by commercial breaks and the network insistence upon its "irresponsibility" for the views and news presented (*presentation is not representation*) we forget that the networks belong to the very global order whose local troubles are our concern.

Nothing pulls us into this global order so much as the vision of its end—an electrical event inscribed on everyone's memory by the techno-mushroom cloud of Hiroshima where Americans permanently scarred themselves with the horror of being obliterated by their own "better" bomb. Ever since, we have been indoctrinated to contemplate the unimaginable drop on America. Thus ABC's *The Day After*, which was shown to millions of viewers in November 1983, actually pictured the survival of several families, along with a fragment of the medical corps and a mid-West campus, after a hair-raising atomic blast which somehow did not melt down the very possibility of such reportage, let alone the technological renaissance that is assigned to these pioneers of cave culture in post-nuclear America. By contrast, when Terri Nash, from Canada's National Film Board, made *If You Love This Planet* (1982) in a 26-minute documentary using Japanese film on the horrors of the human and environmental effects of a single bomb, it was confined to a single national showing on the night it won an Academy Award. As of 1987, the U.S. Justice Department requires registration of screenings and persons viewing the film which it classifies as "political propaganda." The difference between the two films is one that nuclear TV cannot stand, namely, that in one case we see the actual horror of a bomb dropped by Americans and that, in the other case, we see the fictional horror of a bomb imagined by Americans to have been dropped upon Americans by someone else. Thus TV is engaged in a retrospective political history on the nuclear front which parallels the remaking of the Vietnam "War" and America's rehumanization of military adventure.[3]

In the last analysis, television is the blind spot in a culture that for the first time in human history speculates upon the survival of its own rebarbarization. Nothing challenges our humanity as much as this extraordinary presumption of modern cave culture. Until now this sly earth has repeated itself from age to age and the cycle of life has run continuously from the first human beings to ourselves. *Today, however, it is we who must consider ourselves alien.* This is so because we are the first human beings who have to consider seriously the possibility that we may be the last of mankind. This is a possibility that in my opinion renders us the least of mankind and not at all the greatest, as we like to think while contemplating the power of our industry and technology. We have reached this impasse not from any defect in our technology nor from the inability of our immediate environment to sustain the ecological imbalances with which we threaten our biosphere.[4] These risks are, of course, considerable. But the distance between ourselves and the first mankind arises from the power we now hold over nature and ourselves to put an end to all forms of life. The monstrosity of this potential act of ours is beyond comparison if not comprehension. For we are mistaken to think that the extinction of the biosphere is like the extinction of any plant or animal form, or like the obsolescence of an automobile, a dress fad, or the passing of a political regime or art style. This is because the record of such evolutions is due to our practice of historical and scientific reconstruction of the past through which we give to nature and ourselves an identity and prospective vision even in the (un)natural history of the dinosaurs of Jurassic Park.[5] These histories, then, are no more outside of us than we are outside of them. Moreover, it is now highly dangerous to think so, inasmuch as such a practice invites us to imagine ourselves as the surviving historians of the end of the world, whereas we shall be the unrecorded victims of total extinction.

Here is how the last judgment appears to Jonathan Schell, whose reflections upon our ability to annihilate ourselves in a single nuclear holocaust set the most urgent vision of our ecological crisis:

> A nuclear holocaust, because of its unique combination of immensity and suddenness, is a threat without parallel; yet at the same time it is only one of countless threats that the human enterprise, grown mighty through knowledge, poses to the natural world. Our species is caught in the same tightening net of technical success that has already strangled so many other species. (At present, it has been estimated, the earth loses species at the rate of about three per day.) The peril of human extinction, which exists not because every single person in the world would be killed by the immediate explosive and radioactive effects of a holocaust—something that is exceedingly unlikely, even at present levels of armanent—but because a holocaust might render the biosphere unfit for human survival, is, in a word, an *ecological* peril. The nuclear peril is usually seen in isolation from the threats to other forms of life and their ecosystems, but in fact it should be seen as the very center of the ecological crisis—as the cloud-covered Everest of which the more immediate, visible kinds of harm to the environment are the mere foothills. Both the effort to preserve the environment and the effort to save the species from extinction by nuclear arms would be enriched and strengthened by this recognition. The nuclear question, which now stands in eerie seclusion from the rest of life, would gain a context, and the ecological movement, which, in its concern for plants and animals, at times assumes an almost misanthropic posture, as though man were an unwanted intruder in an otherwise unblemished natural world, would gain the humanistic intent that should stand at the heart of its concern.[6]

Modern men, then, are the first among mankind to think of themselves as the last of their kind. In doing so, however, they bring everything else into total darkness. The wholeness of the world which the first men thought of poetically as the body of their body is now truly a body which we can deprive of all life. Our terrible nuclear sun would blind insects, birds, and beasts around the world, destroy the great fishes of the ocean, upset the food chain, render the atmosphere poisonous, scald crops, proliferate cancer, and reduce our civilization to rubble, melting the very electronic circuitry which produced this final and catastrophic flash. The horror of such a prospect cannot be reduced to an individual misfortune, nor to an alien disaster which would leave others of us its horrified but fortunate survivors to pioneer a new society. We are the end of all individualism of this sort. This is so because a nuclear holocaust, like its terrible historical counterpart, is aimed to destroy the very species within which individuals live out expectable births and deaths. The holocaust destroys the very life-world within which individual lives are repeated, renewed, celebrated, and mourned. A nuclear holocaust, as strange as it may seem, is the end of life because it is the end of death;

> Death, having been augmented by human strength, has lost its appointed place in the natural order and become a counter-evolutionary force, capable of destroying in a few years, or even in a few hours, what evolution has built up over billions of years. In doing so, death threatens even itself, since death, after all, is a part of life: stones may be lifeless but they do not die. The question now before the human species, therefore, is whether life or death will prevail on the earth. This is not metaphorical language but a literal description of the present state of affairs.[7]

Whereas the first human beings could think their history and biology within the framework of their own bodies, each resonating in the cosmos of the other, we now entertain the extinction of life and history for all future generations. This is the true extent of our ecological crisis. It is not that we shall leave no forests, no whales, no wild birds, no lakes, no flowers—we shall not. It is not that no tree will shade our graves, nor any flower mark their place, nor even that there will be no one to remember them. We can imagine such sadness even if it is left only to God himself. But what not even God could imagine is that there would be no mankind, no future generation, no life, no birth, no love, no sorrow, no joy, no words, no music, no color in which the Creation might be celebrated from age to age *per omnia saecula saeculorum.*

However did we abrogate to ourselves such a power over nothingness? Even when God punished Adam and Eve with mortality, he did not punish us with the loss of future generations—however painful their birth. And however murderous He found his people to be, not until now could He suffer the murder of all future mankind. Moreover, until now we have faced death, as we have committed murder, with some sense that a life so ended might rest in God's infinite mercy. For we have not believed that nothingness could prevail over God's love for us once He had given us life. Our faith has been that no living creature can entirely abuse the love God first bestowed upon it in the act of creation. The mystery of our being at all seems to us to have set an infinite distance between us and nothingness. We do, of course, imagine ourselves dead, and even others; we murder and commit suicide. But we cannot think the nothingness of never having been born. Such a cry represents the most terrible despair and of all our sins most separates us from God. *By what right, or from what despair, do we then threaten all future generations with nonexistence?*

How dare we so diminish the Creation with the extinction of God's image, with the ruin of cathedrals, of hymns and prayers, with the desolation of this lovely earth in which all things further reflect God? The only thing that could tempt us in such a direction, I believe, is our utilitarianism, by which I mean our prideful assumption that we can use human beings the way we use things. This is to play god with ourselves while lacking divine wisdom as well as divine love. We simultaneously open the door to an infinite abuse of one another over which our secular justice can never prevail, because it, too, has lost its anchorage in the sacred, by which I mean our recognition that life is ultimately outside of our usage so that we set its origin apart from ourselves, as we do God. Here, again, Schell expresses the proper relation between our ultimate values and all other subordinate utilities:

> Human beings have a worth—a worth that is sacred. But it is for human beings that they have that sacred worth, and for them that the other things in the creation have this worth (although it is a reminder of indissoluble connection with the rest of life that many of our needs and desires are also felt by animals). Hence, while our standards of worth have reference to the various possible worthy things in life, they also point back to the life of the needy, or suffering, or rejoicing, or despairing, or admiring, or spiritually thirsting person in whose existence the things are found to be worthy of lacking in worth . . . The death of an individual person is a loss of one subject, and of all needs, longings, sufferings, and enthusiasms—of its being. But the extinction of the species goes farther, and removes from the known universe the human *kind* of being, which is different from any other kind we as yet know of. It is, above all, the death of mankind as an object, that makes extinction radically unique and 'unthinkable'. In extinction a darkness falls over the world not because the lights have gone out but because the eyes that behold the light have been closed.[8]

If we are to respond to the current cultural crisis we must do so by renewing in ourselves the sacred trust in all life which requires us to consecrate future births as the horizon of all human hope and endeavor. If nature is holy it is because it is abundant and its fecundity mirrors what we can hope from God and ourselves. All other practical tasks diminish in the face of our need to restore Life over Death, to hold things in common, to abide with one another and to create a politics and history whose appeal to future generations is not hollowed by nuclear death.

The first men imagined any violation of the reciprocities between their gods, nature, and community to be like a violation of the body or of the family. They therefore made murder and incest things of horror. Today we have difficulty in associating any such horror with the violation of our bodies, or our families and sexuality. We are, of course, concerned with murder, rape, and pornography, as well as with every form of family and child abuse. We sense that the pollution of the family environment must also be counted in our ecological crisis. Young people, whose cynicism amazes us only because it amplifies our own, sense that we no longer invest birth, marriage, and death with the necessary sacredness to commit them in turn to the love of future generations. They have been accused of a natural selfishness or else of willful abdication from the burden of future civilization. These are contradictory accusations and they are dangerous because they weaken the will of their elders to see themselves continued in their children. This in turn renders the elders more prone to use their power in their self-destruction of our very species.

Nevertheless, most of us continue our ordinary lives with unremarkable care for our homes, our floors and dishes, our gardens, our flowers and plants. We care for our cities and streets, for our parks and monuments. We look after ourselves, our children, and our pets, and we care for our friends and neighbors. We do so from day to day, year in and year out, from generation to generation. We find it hard to believe that anyone can trash this earth, or brutalize its people, or destroy themselves. We know they do. But we resist its law and we remain unconvinced of its explanations. We prefer rather to recognize in ourselves a debt of love given to us with this splendid earth and with this life which is ultimately no more ours than the stars in the distant sky. Because of this we count all things sacred, despite our necessary uses of them which oblige us to do so well and with a care for their replacement.[9] And we are richer in this the more we contemplate the legacy of past generations whose culture and civilization have pleased our eyes and our ears, have welcomed us and warmed us without ever knowing us.

All this is desecrated by any thought of the death of the earth or of the extinction of our kind. Such possibilities introduce the outer darkness and bring nothingness where the lamps of civilization and human continuity have flickered unfailingly. In this light we are our own small gods and our temples praise us so long as we venerate life, mournful of its margins of death but with an invincible love of future generations in whom these same words will be made flesh.

NOTES

1. Joyce Nelson, *The Perfect-Machine: TV in the Nuclear Age* (Toronto: Between the Lines, 1987).

2. *Ibid.*, pp. 177-178; Daniel C. Hallin, "Network News: We Keep America on Top of the World," in *Watching Television*, Todd Gitlin, ed. (New York: Pantheon Books, 1987), pp. 9-41.

3. Michael Ryan and Douglas Kellner, *Camera Politica: The Politics and Ideology of Contemporary Hollywood Film* (Bloomington: Indiana University Press, 1988), Chapter 7, "Vietnam and the New Militarism."

4. Preston Cloud, "The Biosphere," *Scientific American*, 249:3 (September, 1983):176-189.

5. John O'Neill, "Dinosaurs-R-Us: The (Un) Natural History of Jurassic Park, in *Monster Theory: Reading Culture*, Jeffrey Jerome Cohen, ed. (Minneapolis: University of Minnesota Press, 1998), pp. 292-308.

6. Jonathan Schell, *The Fate of the Earth* (New York: Avon Books, 1982), p. 111; also, Richard P. Turco, Owen S. Toon, Thomas P. Ackerman, James B. Pollack, and Carl Sagan, "The Climatic Effects of Nuclear War," *Scientific American*, 251:2 (August 1984):33-43.

7. *The Fate of the Earth*, p. 113.

8. *Ibid.*, pp. 127-128.

9. See my essay on the care of the world that we owe one another in *The Communicative Body*, Part Two.

Author Index

Subject Index